Martin Gummert • Nguyen Van Hung
Pauline Chivenge • Boru Douthwaite
Editors

Sustainable Rice Straw Management

Editors
Martin Gummert
Mechanization and Postharvest Cluster, Sustainable Impact Platform
International Rice Research Institute
Los Baños, Laguna, Philippines

Nguyen Van Hung
Mechanization and Postharvest Cluster, Sustainable Impact Platform
International Rice Research Institute
Los Baños, Laguna, Philippines

Pauline Chivenge
Soil, climate, and water Cluster, Sustainable Impact Platform
International Rice Research Institute
Los Baños, Laguna, Philippines

Boru Douthwaite
Selkie Consulting Ltd.
Westport, Mayo, Ireland

https://doi.org/10.1007/978-3-030-32373-8

Preface

Labor shortages in the agricultural sector in South and Southeast Asia associated with rapid economic, social, and political changes have led to accelerated mechanization, particularly involving combine harvesters in rice-based cropping systems. Compared to traditional harvesting methods, combine harvesters leave rice straw on the field. The intensification of cropping systems is also resulting in a larger volume of rice straw being produced that, in turn, must be managed over shorter turnaround times between crops. Both of these trends have led to an increase in open-field burning of the straw because it is the easiest option for farmers. While open-field burning can have positive effects on managing pests, it leads to loss of nutrients and creates air pollution that causes human respiratory ailments. In 2018, burning of rice straw and other agricultural residues—which contributes to poor air quality—prompted the Indian government to ban open fires in New Delhi. Burning the straw also removes opportunities for adding value to it.

More sustainable rice straw management methods are urgently needed to minimize rice production's carbon footprint and its negative effects on human health and to maximize adding value to the straw byproduct. Past research on rice straw has focused on isolated topics or component technologies, e.g., to improve straw combustion properties or to analyze nutritional value as an animal feedstock. However, to date, there has been no holistic approach toward rice straw research. Topics that should be considered include (1) the effects of burning straw, (2) incorporation or removal of straw from the field, (3) processing and utilizing straw as different agricultural or industrial byproducts, (4) determining the impact of straw on soil fertility and nutrient cycling, (5) assessing the impact of greenhouse gas (GHG) emissions and environmental pollution on human health, and (6) developing economic value-adding opportunities.

Until recently, applied and science-based data had not been available. In fact, when the International Rice Research Institute (IRRI) started its rice straw research, no one could define exactly what constitutes sustainable rice straw management. So, in 2012, IRRI initiated a research program on sustainable rice straw management by hosting an international workshop on rice straw energy. This was followed by projects involving a feasibility study on a rice straw combustion plant using organic rankine cycle technology (2013–2014) and rice straw bioenergy (2013–2016).

Based on this initial work, a BMZ-funded project was implemented during 2016–2018 on *Scalable straw management options for improved farmer livelihoods, sustainability, and low environmental footprint in rice-based production systems*. Its objectives were to (1) identify, develop, and verify technologies and business models for sustainable rice straw management; (2) conduct market studies on existing and potential rice straw product markets; (3) establish data on GHG emissions from different rice straw management and processing practices; (4) determine environmental footprints using life cycle assessment (LCA); and (5) formulate policy recommendations for communicating to policy makers.

This book summarizes, in part, the outputs of the above-mentioned projects by IRRI and its national agricultural research and extension partners in Vietnam and the Philippines. It also includes complementary contributions from other experts on selected topics that were not covered by the IRRI projects. The book is aimed at engineers and researchers interested in current good practices and the gaps and constraints that require further research and innovation.

By no means an end in itself, this book provides an overview of research activities on straw management in the two countries. It basically provides a snapshot of what we know and have learned through to the completion of the workshops and projects during 2012–2018. This accumulated information can be used to help farmers and extension workers decide on the best alternative straw management options by presenting technological options, as well as the value chains and business models required to make them work. Finally, the book provides research-based evidence that may guide policy makers in South and Southeast Asia—required by the public to reduce GHG emissions and air pollution—to develop and implement appropriate policies. See the table of contents for the list of topics in the 11 chapters. More research is needed on (1) the long-term effects on soil fertility, (2) the effects on the environment and health caused by changed on-field rice straw management, (3) more complex rice straw products with more value-adding potential, (4) second-generation bio fuels and bio refineries, (5) and the sustainability of the various options. In addition, research studies that focus on the trade-offs and synergies of different straw management options remain pertinent.

Los Baños, Laguna, Philippines Martin Gummert
Los Baños, Laguna, Philippines Nguyen Van Hung
Los Baños, Laguna, Philippines Pauline Chivenge
Westport, Mayo, Ireland Boru Douthwaite

Acknowledgments

We would like to acknowledge the financial support of (1) the German Federal Ministry for Economic Cooperation and Development (BMZ) through the project *Scalable straw management options for improved livelihoods, sustainability, and low environmental footprint in rice-based production systems* (Grant Ref: 15.7860.8-001.00); (2) the Philippine Department of Agriculture through the project *Sustainable rice straw management for bioenergy, food, and feed in the Philippines* (Grant ref: 101101-03-042-2018); (3) Flagship Program 2, Upgrading Rice Value Chains of the CGIAR Research Program on Rice (RICE) and (4) the Swiss Agency for Development and Cooperation (SDC) through the CORIGAP project [Grant no. 81016734].

The authors acknowledge the valuable support of the management and research support teams at the International Rice Research Institute.

Contents

Contributors

Justin Allen International Rice Research Institute (IRRI), Los Baños, Laguna, Philippines

Daniel Aquino Philippine Carabao Center, Central Luzon State University (CLSU), Muñoz, Philippines

Carlito Balingbing International Rice Research Institute (IRRI), Los Baños, Laguna, Philippines

Elmer Bautista Philippine Rice Research Institute (PhilRice), Los Baños, Philippines

Francis Mervin S. Chan University of the Philippines Los Baños, Los Baños, Philippines

Pauline Chivenge International Rice Research Institute (IRRI), Los Baños, Laguna, Philippines

Rizal G. Corales Philippine Rice Research Institute (PhilRice), Los Baños, Philippines

Do Minh Cuong Hue University, Hue, Vietnam

Marie Claire Custodio International Rice Research Institute (IRRI), Los Baños, Laguna, Philippines

Arnel Del Barrio Philippine Carabao Center, National Headquarters, Muñoz, Philippines

Matty Demont International Rice Research Institute (IRRI), Los Baños, Laguna, Philippines

Nguyen Xuan Du Sai Gon University, Ho Chi Minh City, Vietnam

Giang Phuong Duong Food and Agriculture Organization of the United Nations (FAO), Rome, Italy

Martin Gummert International Rice Research Institute (IRRI), Los Baños, Laguna, Philippines

Hinh The Nguyen Nong Lam University, Ho Chi Minh City, Vietnam

Phan Hieu Hien Nong Lam University, Ho Chi Minh City, Vietnam

Nguyen The Hinh Ministry of Agriculture and Rural Development (MARD), Hanoi, Vietnam

Dinh Vuong Hung Hue University, Hue, Vietnam

Nguyen Van Hung International Rice Research Institute (IRRI), Los Baños, Laguna, Philippines

Duong Nguyen Khang Nong Lam University, Ho Chi Minh City, Vietnam

Vu Tien Khang Cuu Long Rice Research Institute, Can Tho City, Vietnam

Monet Concepcion Maguyon-Detras Department of Chemical Engineering, College of Engineering and Agro-Industrial Technology, University of the Philippines Los Baños (UPLB), Los Baños, Laguna, Philippines

Maria Victoria Migo Department of Chemical Engineering, College of Engineering and Agro-Industrial Technology, University of the Philippines Los Baños (UPLB), Los Baños, Laguna, Philippines

Tran Sy Nam Can Tho University, Can Tho, Vietnam

Nguyen Vo Chau Ngan Can Tho University, Can Tho, Vietnam

Nguyen Thanh Nghi Nong Lam University, Ho Chi Minh City, Vietnam

Ninh Thai Hoang Ministry of Agriculture and Rural Development (MARD), Hanoi, Vietnam

Virgina Ompad Philippine Rice Research Institute (PhilRice), Los Baños, Philippines

Kristine S. Pascual Philippine Rice Research Institute (PhilRice), Muñoz, Nueva Ecija, Philippines

Reianne Quilloy International Rice Research Institute (IRRI), Los Baños, Laguna, Philippines

Remelyn E. Ramos Philippine Rice Research Institute (PhilRice), Los Baños, Philippines

Ryan R. Romasanta International Rice Research Institute (IRRI), Los Baños, Laguna, Philippines

Ampy Paulo Roxas International Rice Research Institute (IRRI), Los Baños, Laguna, Philippines

Francis Rubianes International Rice Research Institute (IRRI), Los Baños, Laguna, Philippines

Julius T. Sajor Philippine Rice Research Institute (PhilRice), Los Baños, Philippines

Bjoern Ole Sander International Rice Research Institute (IRRI), Tu Liem District, Hanoi, Vietnam

Dang Thanh Son University of Technical Education, Vinh Long, Vietnam

Caesar Joventino Tado Philippine Rice Research Institute (PhilRice), Los Baños, Philippines

Le Vinh Thuc Can Tho University, Can Tho, Vietnam

Toàn Minh Dương Can Tho University, Can Tho, Vietnam

Nguyen Tat Toan Nong Lam University, Ho Chi Minh City, Vietnam

Nguyen Xuan Trach Vietnam National University of Agriculture, Hanoi, Vietnam

Thi Thanh Truc Ngo Can Tho University, Can Tho, Vietnam

Duong Van Chin Dinh Thanh Agricultural Research Center, Loc Troi Group, Hoa Tan Hamlet, Thoai Son District, An Giang, Vietnam

Nguyen Van Hieu Tien Giang University, Tien Giang, Vietnam

Tran Van Thach Dinh Thanh Agricultural Research Center (DTARC), Loc Troi Group, Hoa Tan Hamlet, Thoai Son District, An Giang, Vietnam

Huynh Van Thao Can Tho University, Can Tho, Vietnam

Mai Van Trinh Institute of Agricultural Environment, Nam Tu Liem District, Hanoi, Vietnam

Le Quang Vinh Nong Lam University, Ho Chi Minh City, Vietnam

Chapter 1
Rice Straw Overview: Availability, Properties, and Management Practices

Nguyen Van Hung, Monet Concepcion Maguyon-Detras, Maria Victoria Migo, Reianne Quilloy, Carlito Balingbing, Pauline Chivenge, and Martin Gummert

Abstract Managing rice straw remains a challenge in Asia where more rice, and hence, more straw, is grown each year to meet rising demand. The widespread burning of rice straw is a major contributor to dangerously high levels of air pollution in South- and Southeast Asia associated with health issues. At the same time, researchers, engineers, and entrepreneurs are developing a range of alternative uses that turn rice straw into a commodity around which sustainable value chains can be built to benefit rural people. The best alternative to burning rice straw in any one location depends on context. However, available information remains scattered in different media and no publication yet exists that helps people learn about, and decide between, rice straw management options. This book provides a synthesis of these options and integrates knowledge on relevant areas: sustainable rice straw management practices, rice straw value chains, and business models. The book is also based on new research and practice data from research organizations and innovators in Vietnam, the Philippines, and Cambodia.

Keywords Rice · Rice straw · Residue · Sustainable · Rice straw management

N. V. Hung (✉) · R. Quilloy · C. Balingbing · P. Chivenge · M. Gummert
International Rice Research Institute (IRRI), Los Baños, Laguna, Philippines
e-mail: hung.nguyen@irri.org; r.quilloy@irri.org; c.balingbing@irri.org; p.chivenge@irri.org; m.gummert@irri.org

M. C. Maguyon-Detras · M. V. Migo
Department of Chemical Engineering, College of Engineering and Agro-Industrial Technology, University of the Philippines Los Baños (UPLB), Los Baños, Laguna, Philippines
e-mail: mmdetras@up.edu.ph; mpmigo@up.edu.ph

M. Gummert et al. (eds.), *Sustainable Rice Straw Management*,
https://doi.org/10.1007/978-3-030-32373-8_1

1.1 Rice Straw Availability

Rice straw is a residual byproduct of rice production at harvest. The total biomass of this residue depends on various factors such as varieties, soils and nutrient management and weather. At harvest, rice straw is piled or spread in the field depending on the harvesting methods, using stationary threshers or self-propelled combine harvesters, respectively. The amount of rice straw taken off the field depends mainly on the cutting height (i.e., height of the stubble left in the field). Rice straw that remains in the field after harvest can be collected, burned, or left to decompose (soil incorporation). The "stubble"—the uncut portion of the rice straw after harvest—remains, and can be burned or incorporated into the soil in preparation for the next crop. The ratio of straw to paddy varies, ranging from 1.0 to 4.3 (Zafar 2015) and 0.74–0.79 (Nguyen-Hung et al. 2016a). We investigated biomass ratios for a common rice variety (NSIC Rc158) at IRRI in 2017 that resulted in the findings shown in Fig. 1.1 (unpublished). Yield of the total straw biomass ranges from 7.5 to 8 t/ha while removed straw (harvested with leftover grains) ranged from 2.7 to 8 t/ha corresponding to the cut portion ranging from 50% to 100% of the total straw biomass. Figure 1.2 shows the global minimum and maximum estimate of rice straw avail-

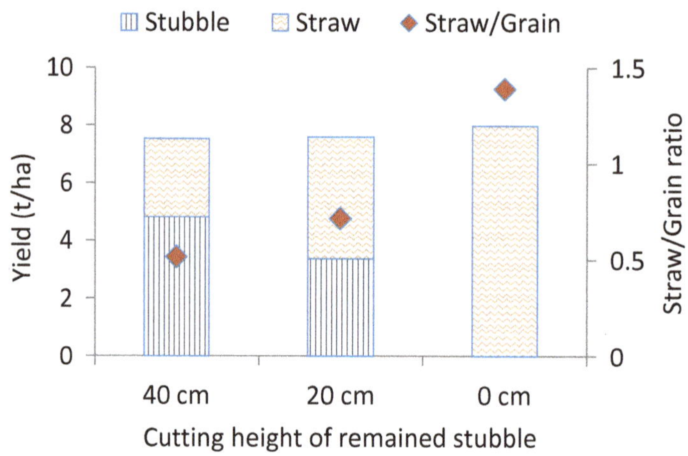

Fig. 1.1 Biomass ratios of rice production for the NSIC Rc158 variety grown at IRRI during the 2017 dry season

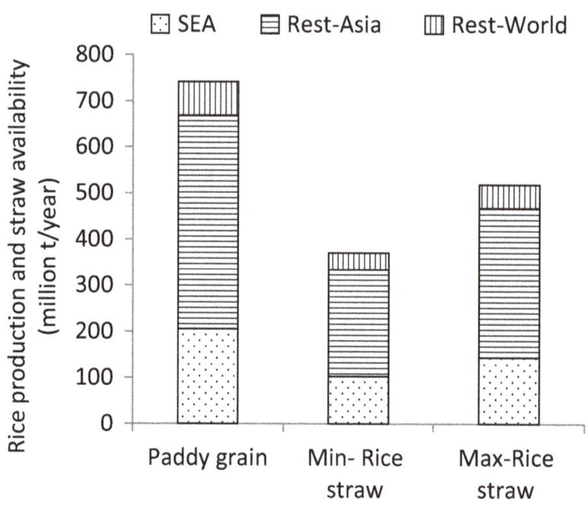

Fig. 1.2 Annual Rice production and rice straw availability in SEA, the rest of Asia, and the rest of the World

ability based on global rice production data (IRRI 2019) and the straw:grain ratios of 0.5 and 0.7 from the experiment.

Annual rice straw production is in the ranges of 100–140, 330–470, and 370–520 million t/year in Southeast Asia (SEA), the whole of Asia, and over the world, respectively (Fig. 1.2).

1.2 Rice Straw Properties and Composition

Utilization of rice straw is dependent on its characteristics, which can be divided into three major categories: (1) physical properties, (2) thermal properties, and (3) chemical composition. Physical properties include bulk density, heat capacity, and thermal conductivity. Density is the most relevant to the handling and storage of rice straw. Thermal properties, and heating value; these properties are relevant when biomass is converted to energy. Chemical composition, such as lignin, cellulose, hemicellulose/carbohydrates, and nutrient contents, are relevant to applications, such as for livestock feed and soil fertility. Characterizing rice straw is helpful for life cycle analysis and efficiency calculations. The most common methods used in the characterization of rice straw can be referenced from the National Renewable Energy Laboratory (NREL) and the American Society for Testing and Materials (ASTM).

1.2.1 Physical Properties

Based on various studies, the bulk density of rice straw can vary depending on the different forms it may take. Loose rice straw, collected directly from the field, can range in density from 13 to 18 kg m^{-3} in dry matter (dm) (Migo 2019). Chopped straw, ranging in length from 2 to 10 mm (Chou et al. 2009), can have a density range of from 50 to 120 kg m^{-3} (Liu et al. 2011), depending on the equipment used. Depending on the baler equipment used, baled straw size and the compression ratio, and thus bulk density, will vary. A round rice straw bale with a 70-cm length and 50-cm diameter has a bulk density ranging from 60 to 90 kg m^{-3} dm (Nguyen-Van-Hung et al. 2016b). The density of rice straw briquettes with a 90-mm diameter and 7- to 15-mm thickness is 350–450 kg m^{-3} dm (Munder 2013). The density of rice straw pellets with an 8-mm diameter and from 30 to 50 mm in height is 600–700 kg m^{-3} dm (Nguyen-Van-Hieu et al. 2018).

As compared to rice husks, which have a density of between 86 and 114 kg m^{-3} (Mansarav and Ghaly 1997), unprocessed, loose rice straw has a low density. This means a higher volume per kilogram, implying higher shipping and handling costs as well as more complications in processing, transportation, storage, and burning (Duan et al. 2015, Liu et al. 2011). Rice straw volume can be reduced through processing but this will require additional energy inputs. Various size-reduction methods can increase density of the straw including using of pellet mills (Nguyen-V-Hieu et al. 2018), roller presses, piston presses, cubers, briquette presses, screw extruders, tabletizers, and agglomerators (Satlewal et al. 2017).

When used for bioenergy, rice straw's bulk density influences the combustion process as it affects the time required in the reactor (Zhang et al. 2012). Rozainee et al. (2008), as cited by Zhang et al. (2012), reported that a low bulk density causes poor mixing and nonuniform temperature distribution (unfavorable operating conditions), which decreases energy efficiency.

The *moisture content* of rice straw is an important consideration when determining how to process it and what it will be used for. For example, moisture content affects the heating value of the straw, which is important when the byproduct is intended for use as bioenergy. In addition, if rice straw volume is to be reduced, the moisture content before compression should be between 12 and 17% (Kargbo et al. 2010). Unfortunately, the moisture content can fluctuate greatly due to the method and duration of the straw's storage (Topno 2015).

1.2.2 Thermal Properties

The *calorific value* is an essential parameter that shows the energy value of rice straw, if to be used for bioenergy. Rice straw's energy efficiency can be calculated by dividing its energy output by its calorific value, which may be expressed as the higher-heating value (HHV), wherein latent heat of the water is included, or lower-heating value (LHV). In terms of calorific value, rice straw has an HHV that ranges from 14.08 to 15.09 MJ kg^{-1}, as determined by different studies as shown in Table 1.1 and is comparable to rice husks with a calorific value of around 14.2 MJ kg^{-1}. However, the calorific value of rice straw is just one-third of that of kerosene, which has a calorific value of 46.2 MJ kg^{-1}.

In the *proximate analysis*, volatiles refer to the volatile carbon, combined water, net hydrogen, nitrogen, and sulfur, which are first driven off in combustion. Rice straw is characterized by high volatiles or volatile matter (VOM) (60.55–69.70%), which is comparable to the biomass of other byproducts, such as sugar cane bagasse, corn straw, wheat straw, etc. In bioenergy applications, specifically in combustion, a high VOM has advantages, such as easier ignition and burning; but it also leads to a rapid, more difficult-to-control combustion (Liu et al. 2011). Fixed carbon refers to the carbon left after the volatiles are driven off. Rice straw has a fixed carbon ranging from 11.10% to 16.75%, which is also comparable to other biomass.

The *ultimate analysis* reveals the elemental carbon, hydrogen, oxygen, nitrogen, and sulfur composition of rice straw. Compared to fossil fuels, the carbon content of rice straw biomass is less, while the oxygen and hydrogen contents are higher. As shown in Fig. 1.3, the van Krevelen diagram shows the hydrogen-to-carbon (H:C) and oxygen-to-carbon (O:C) ratios of various fuels. The ranges of H:C and O:C in rice straw are 1.1–1.36 and 0.94–1.06, respectively, which place it in the biomass region of the van Krevelen diagram, specifically in the cellulose region.

Rice straw *ash content*, which includes noncombustible residues, is around 18.67–29.1%. The high silica content of rice straw (Table 1.2) causes erosion prob-

Table 1.1 Calorific value and proximate and ultimate analyses of rice straw

	HHV MJ/kg	Proximate analysis (% dry fuel)		Ultimate analysis (% dry fuel)								Sources
		Fix C	Volatiles	Ash	C	H	O	N	S	Cl	Ash	
	15.09	15.86	65.47	18.67	38.24	5.2	36.26	0.87	0.18	0.58	18.67	Jenkins et al. (1996)
		11.10	69.70	19.20								Braunbeck (1998)
	14.57				35.94			1.18			22.00	Munder (2013)
	14.08				33.70	4.0		1.71	0.16	0.32	29.10	Guillemot et al. (2014)
	15.03	13.21	64.24	13.26	44.40	7.40	47.07	1.13				Duan et al. (2015)
	14.39	16.75	60.55	22.70	35.35	3.91	37.35	0.71	0.03			Migo (2019)
Range	14.08	11.10	60.55	13.26	33.70	3.91	36.26	0.71	0.03	0.32	18.67	
	−15.09	−16.75	−69.70	−22.70	−44.40	−7.40	−47.07	−1.71	−0.18	−0.58	−29.10	

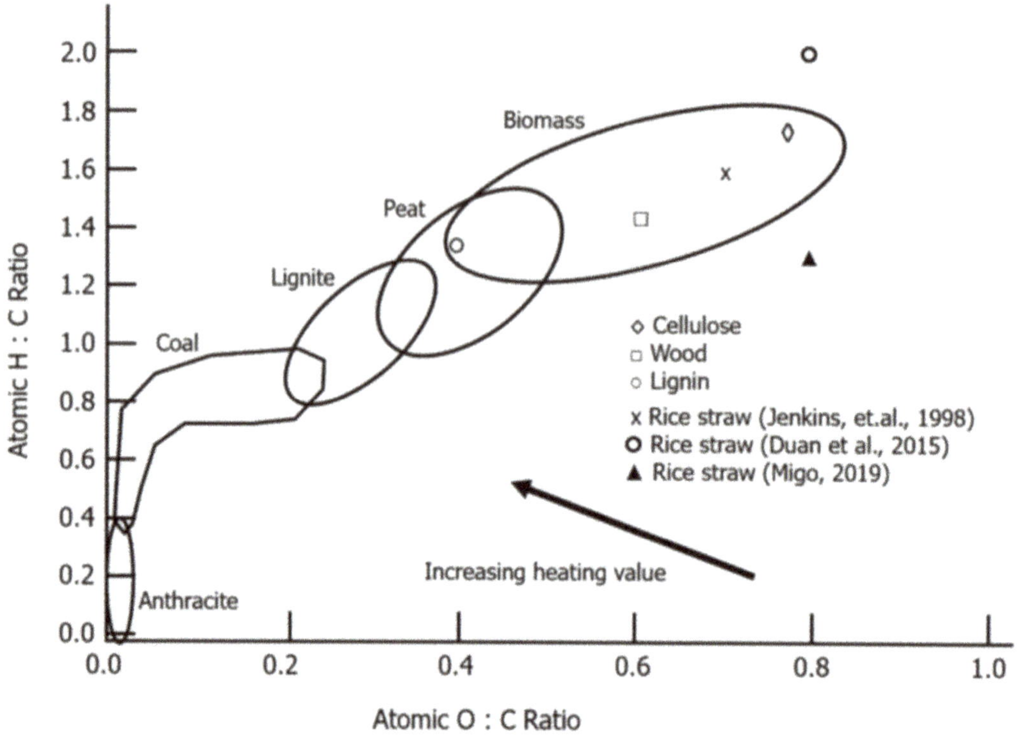

Fig. 1.3 Van Krevelen diagram for various solid fuels. Source: Adapted from Mando (2013)

Table 1.2 Rice straw ash properties

	SiO_2	Al_2O_3	TiO_2	Fe_2O_3	CaO	MgO	Na_2O	K_2O_3	SO_3	P_2O_5	Sources
% of ash (d.b)	75.00	1.40	0.02	2.00	1.50	1.90	1.90	10.00	0.90	2.70	Liu, et al. (2011)
	74.67	1.04	0.09	0.85	3.01	1.75	0.96	12.30	1.24	1.41	Jeng, et al. (2012)
	82.60	1.10	0.60	1.00	3.30	1.70	0.30	6.30	0.90	1.70	Guillemot (2014)
	67.78	1.54			2.08	1.11	1.48	11.87			Migo (2019)
Range	67.78	1.04	0.02	0.85	2.08	1.11	0.30	6.30	0.90	1.41	
	−82.60	−1.54	−0.6	−2.00	−3.01	−1.90	−1.90	−12.30	−1.24	−2.70	

lems in processing machines (for example, in conveyers and grinders), boilers, and decreases the digestibility of rice straw when used as fodder. Rice straw is also characterized by a high volatile matter as compared to wood and coal; and a lower fixed carbon compared than that in coal. The high ash content in rice straw decreases its calorific value and causes problems in energy conversion. A high potassium and alkali content in ash may increase corrosion and fouling problems in grates, since alkali metals are known triggers for these phenomena. Table 1.3 shows the ash analysis of rice straw.

Table 1.3 Chemical composition of rice straw

	DM	CP											
	%	% DM	Crude fiber	NDF	ADF	ADL	EBSi	Ash	Ca	P	Na	K	Sources
	92.8	4.2	35.1	69.1	42.4	4.8		18.1	0.29	0.09	0.27	1.8	Ngi, et al. (2006)
	96.3			73.0	41.6	4.8	4.3	12.1	1.58	0.12	0.13	3.4	Sarnklong et al. (2010)
	90.6	4.2		73.2	44.9	3.2							Peripolli et al. (2016)
Range	90.6	4.2	35.1	69.1	41.9	3.2	4.3	12.1	0.29	0.09	0.13	1.8	
	−96.3			−73.2	−44.9	−4.8		−18.1	−1.58	−0.12	−0.27	−3.4	

DM dry matter, *CP* crude protein, *NDF* neutral detergent fiber, *ADF* acid detergent fiber, *ADL* acid detergent lignin, *EBSi* extractable biogenic silica

1.2.3 Chemical Composition

Chemical composition determines the nutritional quality of rice straw, which is important for livestock feed, anaerobic digestion, and as a soil amendment. Rice straw has low nutritional value and research has been done to improve it. Jenkins (1998) indicated that the typical components of plant biomass are moisture cellulose, hemicelluloses, lignin, lipids, proteins, simple sugars, starches, water, hydrocarbon, ash, and other compounds. The concentrations of these compounds depend on the plant species, type of tissue, growth stage, and growing conditions. Rice straw is considered a lignocellulosic biomass that contains 38% cellulose, 25% hemicellulose, and 12% lignin (Japan Institute of Energy 2002). Compared to the biomass of other plants, such as softwood, rice straw is lower in cellulose and lignin and higher in hemicellulose content (Barmina et al. 2013). Table 1.3 shows the compositional analysis of rice straw via the work of various researchers.

1.3 Overview of Rice-Straw Management Options

1.3.1 Burning Issues and Alternative Management Options

Intensification of rice-cropping systems has been associated with the use of high-yielding and short-duration varieties with shorter turnaround time between crops in multi-cropping systems. Furthermore, the rapid introduction of combine harvesters constitutes a game changer because of the larger amounts of straw that are left spread out on the field. Manual collection of the straw in the field is unprofitable because of the high labor cost. Incorporation in the soil poses challenges in intensive systems with two to three cropping rounds per year. This is due to the insufficient time for decomposition, leaving the straw with poor fertilization properties for the soil and hindering crop establishment. As a result, open-field burning of straw has increased dramatically over the last decade, despite being banned in most rice-growing countries because of pollution and the associated health issues. Therefore, it is important to look for sustainable solutions and technologies that can reduce the environmental footprint and add value by increasing the revenues of rice production systems. Options for rice-straw management are shown in Fig. 1.4. Rice straw can inherently be used for soil conditioning thru composting and carbonization; as well as for bio-energy production and for materials recovery such as silica and bio-fiber (for industrial use). It is important to note that not all the possible options are economically viable. This is due to the fact that the processing material and transportation costs in value-adding solutions are still higher as compared to using the other more traditional options.

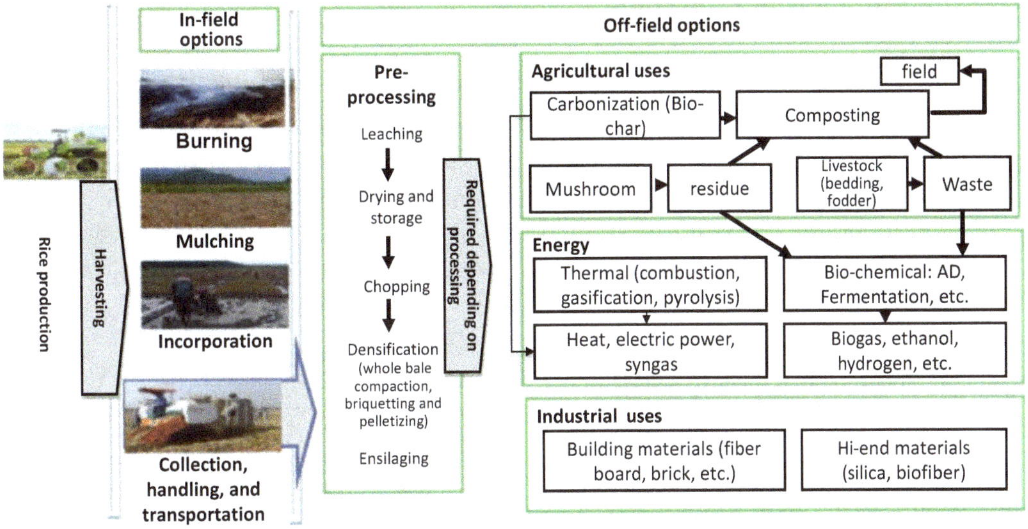

Fig. 1.4 Rice-straw management options

1.3.2 Scalable Solutions for Sustainable Rice-Straw Management

1.3.2.1 Incorporation

Rice straw incorporation into soil is another common management option, but adequate time must be allowed for its decomposition to ensure effectiveness and production efficiency (Mandal et al. 2004; Yadvinder-Singh et al. 2004; Dobermann and Fairhurst 2002). Additionally, careful straw management considerations have to be made after soil incorporation for greenhouse gas emission (GHGE) (Sander et al. 2014). Rice straw is characterized by a slow decomposition rate; thus, some farmers avoid rice straw soil incorporation especially in intensive cropping systems with 3 weeks interlude. In terms of total carbon dioxide equivalent (CO_2-eq) per ha converted from CH_4 and N_2O, recent researches at IRRI showed that rice straw soil incorporation emitted about from 3500 to 4500 kg CO_2-eq ha^{-1} (Rosamanta 2017) which is about 1.5–2.0 times higher than when rice straw was removed. In response to this, researchers have conducted studies to evaluate using fungal inocula to speed up the decomposition rate (Goyal and Sindhu 2011, Ngo-T-T-Truc et al. 2012). Rice straw is chopped with combine harvesters and then sprayed with an inoculum to foster its decomposition in the soil. This management option is discussed in more detail in Chap. 9.

1.3.2.2 Mechanized Collection

Combine harvesters are known to spread rice straw across the field. Therefore, since rice straw collection is energy intensive, it is only economically viable and practical thru mechanical collection by use of balers. Collection plays a critical role in the rice straw supply chain. A discussion on different rice straw balers used in Asia is presented in Nguyen-V-Hung et al. (2017). Mechanized collection technologies are discussed in more detail in Chap. 2.

1.3.2.3 Mechanized Composting

Rice straw composting is done by adding animal manure and enzymes to rice straw and mixing by a turner and ensilage, in order to homogenize the mixture. The bio-physical processes of decaying matter can drastically improve thru mechanized composting. In turn, the compost can serve as fertilizer for growing vegetables and other crops, or can be used directly as soil conditioner. As soil conditioner, it improves the nutrient and organic matter content of the soil. This technology is described in more detail in Chap. 3.

1.3.2.4 Mushroom Production

The species of rice-straw mushrooms, *Volvariella volvacea,* is commonly used because of it grows easily and has a short growth duration of 14 days. The species grows in tropical weather at around 30–35 °C for the mycelia development stage, and at around 28–30 °C for the fruiting body production stage. The main inputs for mushroom growing are rice straw, spawn, labor, and water. The mushroom harvest usually starts during the third week after inoculation and ends 1 week later. Outdoor mushroom production is a common practice in Vietnam's Mekong River Delta (MRD). The low investment cost is an advantage of this income-generating enterprise. It produces a yield of 0.8 kg of mushrooms per 10 kg of dried straw and generates a net profit of USD 50–100 t^{-1} of straw. Indoor production is a less common practice because of higher investment costs and the necessary strict control of the growing conditions. On the other hand, indoor mushroom growing produces about a 2-kg higher yield per 10 kg of dried straw. See Chap. 7 for more details on mushroom production.

1.3.2.5 Rice-Straw Silage for Cattle Feed

Rice straw is of poor quality to serve as a livestock feed. It has a low C:N ratio and high NDF and ADF, which affects its nutritive value. Nevertheless, it is considered as a potential feed additive for increasing the energy and protein content. The prescribed consumption limit of rice straw by ruminants is 1.0 to 1.5 kg per 100 kg live-weight per day (Drake et al. 2002). Urea treatment of straw, which is rice straw ensilaged with 2–4% urea can improve consumption and digestibility of the rice straw as fodder. This technology is discussed in more detail in Chap. 7.

1.4 Conclusions and Recommendations

Upgrading the value chain of rice straw-byproducts and employing sustainable straw-management practices are the key to influencing farmers not to do open-field burning and thus avoid the negative environmental and health consequences. Incorporating rice straw into the soil is an option; however, it needs to be considered carefully to ensure timely decomposition and to minimize GHGE. Mechanized collection with balers plays a critical role in the sustainable use of rice straw. Alternative straw management options, such as straw-based mushroom and feed production, mechanized composting to produce organic fertilizer, etc., are discussed in the remaining chapters of the book.

This book focuses on the scalable options that will add economic value to rice production in Asia. Reviewed and updated information as well as scientific evidence on sustainable rice-straw management will be useful for further developments and related policies. Topics for another publication could be how rice straw can be used to produce biofuel and high-end materials, such as bioplastics, biofibers, and silica.

References

Barmina I, Lickrastina A, Valdmanis R, Zake M, Arshanitsa A, Solodovnik V, Telysheva G (2013) Effects of biomass composition variations on gasification and combustion characteristics. Engineering for rural development. Jelgava 05:23–24

Braunbeck CM (1998) Development of a rice husk furnace for preheating of the drying air of a low temperature drying system, PhD thesis, University of Hohenheim

Chou C, Lin S, Lu W (2009) Preparation and characterization of solid biomass fuel made from rice straw and rice bran. Fuel Process Technol 90(7–8):980–987

Dobermann A, Fairhurst TH (2002) Rice straw management. Better Crops International, vol 16, Special supplement, May 2002. http://www.ipni.net/publication/bci.nsf/0/163087B956D0EFF485257BBA006531E8/$FILE/Better%20Crops%20International%202002-3%20p07.pdf

Drake D, Nader G, Forero L (2002) Feeding rice straw to cattle. Publication 8079. University of California. Division of Agriculture and Natural Resources. http://anrcatalog.ucdavis.edu

Duan F, Chyang C, Zhang L, Yin S (2015) Bed agglomeration characteristics of rice straw combustion in a vortexing fluidized-bed combustor. Bioresour Technol 183:195–202

Goyal S, Sindhu SS (2011) Composting of rice straw using different inocula and analysis of compost quality. Microbiol J 4:126–138

Guillemot A, Bruant R, Pasquiou V, Boucher E (2014) Feasibility study for the implementation of two ORC power plants of 1 MWe each using rice straw as a fuel in the context of a public-private partnership with the institutions PhilRice and UPLB. ENERTIME, Courbevoie

IRRI (International Rice Research Institute) (2019) World rice statistics. http://ricestat.irri.org:8080/wrsv3/entrypoint.htm

Japan Institute of Energy (2002) Asian biomass handbook: a guide for biomass production and utilization. http://www.build-a-biogas-plant.com/PDF/AsianBiomassHandbook2008.pdf

Jeng SL, Zainuddin AM, Sharifah RWA, Haslenda H (2012) A review on utilisation of biomass from rice industry as a source of renewable energy. Renew Sust Energ Rev 16:3084–3094

Jenkins BM (1998) Physical properties of biomass. In: Kitani O, Hall CW (eds) Biomass handbook. Gordon and Breach, New York

Jenkins BM, Bakker RR, Wei JB (1996) On the properties of washed straw. Biomass Bioenergy 10(4):177–200

Kargbo F, Xing J, Zhang Y (2010) Property analysis and pretreatment of rice straw for energy use in grain drying: a review. Agric Biol J N Am 1(3):195–200

Liu Z, Xu A, Long B (2011) Energy from combustion of rice straw: status and challenges to China. Energy Power Eng 3:325–331

Mandal KG, Misra AK, Hati KM, Bandyopadhyay KK, Ghosh PK, Mohanty M (2004) Rice residue management options and effects on soil properties and crop productivity. Food Agric Environ 2:224–231

Mando M (2013) Direct combustion of biomass. Book chapter. In: Biomass combustion science, technology and engineering. Erscheinungsort nicht ermittelbar: Woodhead Publishing

Mansarav KG, Ghaly AE (1997) Physical and thermochemical properties of rice husk. Energy Sources J 19(9). https://doi.org/10.1080/00908319708908904

Migo MVP (2019) Optimization and life cycle assessment of the direct combustion of rice straw using a small scale, stationary grate furnace for heat generation. Unpublished Masters thesis. University of the Philippines Los Baños

Munder S (2013) Improving thermal conversion properties of rice straw by briquetting. Masters thesis. University of Hohenheim

Ngi J, Ayoade JA, Oluremi OIA (2006) Evaluation of dried cassava leaf meal and maize offal as supplement for goats fed rice straw in dry season. Livest Res Rural Dev 18(9). https://www.feedipedia.org/node/3008

Ngo-T-T-Truc, Zenaida MS, Maria VOE, Enrique PP, Corazon LR, Florencia GP (2012) Farmers' awareness and factors affecting adoption of rapid composting in Mekong Delta, Vietnam and Central Luzon, Philippines. J Environ Sci Manag 15(2):59–73

Nguyen-V-Hieu, Nguyen-T-Nghi, Le-Q-Vinh, Le-M-Anh, Nguyen-V-Hung, Gummert M (2018) Developing densified products to reduce transportation costs and improve the quality of rice straw feedstocks for cattle feeding. J Viet Environ 10(1). https://oa.slub-dresden.de/ejournals/jve/article/view/2919

Nguyen-V-Hung, Topno S, Balingbing C, Nguyen-V-C-Ngan, Roder M, Quilty J, Jamieson C, Thornley P, Gummert M (2016a) Generating a positive energy balance from using rice straw for anaerobic digestion. Energy Rep 2:117–122

Nguyen-V-Hung, Nguyen-D-Canh, Tran-V-Tuan, Hau-D-Hoa, Gummert M (2016b) Energy efficiency, greenhouse gas emissions, and cost of rice straw collection in the Mekong River Delta of Vietnam. Field Crops Res 198:16–22

Nguyen-V-Hung, Carlito B, Quilty J, Bojern S, Demont M, Gummert M (2017) Processing rice straw and rice husk as co-products. In: Sasaki T (ed) Achieving sustainable cultivation of rice, vol 2. Burleigh Dodds Science Publishing, Cambridge, pp 121–148

Peripolli V, Barcellos JO, Prates ÊR, Mcmanus C, Silva LP, Stella LA, Lopes RB (2016) Nutritional value of baled rice straw for ruminant feed. Rev Bras Zootec 45(7):392–399

Rozainee M, Ngo SP, Salema AA, Tan KG, Aiffin M et al (2008) Effect of fluidising velocity on the combustion of rice husk in a bench-scale fluidised bed combustor for the production of amorphous rice husk ash. Bioresour Technol 99:703–713

Sander BO, Samson M, Buresh RJ (2014) Methane and nitrous oxide emissions from flooded rice fields as affected by water and straw management between rice crops. Geoderma 235–236:355–362

Sarnklong C, Cone JW, Pellikaan W, Hendriks WH (2010) Utilization of rice straw and different treatments to improve its feed value for ruminants: a review. Asian-Aust J Anim Sci 23(5):680–692

Satlewal A, Agrawal R, Bhagia S, Das P, Ragauskas AJ (2017) Rice straw as a feedstock for biofuels: availability, recalcitrance, and chemical properties. Biofuels Bioprod Biorefin 12(1):83–107

Topno SE (2015) Environmental performance and energy recoverable from stored rice straw bales in humid climate. Unpublished PhD thesis. University of the Philippines Los Baños

Yadvinder-Singh, Bijay-Singh, Ladha JK, Khind CS, Khera TS, Bueno CS (2004) Effects of residue decomposition on productivity and soil fertility in rice–wheat rotation. Soil Sci Soc Am J 68:854–864

Zafar S (2015) Rice straw as bioenergy resource. Bioenergy Consult. http://www.bioenergyconsult.com/tag/rice-residues/

Zhang Y, Ghaly AE, Li B (2012) Physical properties of rice residues as affected by variety and climatic and cultivation conditions in three continents. Am J Appl Sci 9(11):1757–1768

Chapter 2
Mechanized Collection and Densification of Rice Straw

Carlito Balingbing, Nguyen Van Hung, Nguyen Thanh Nghi, Nguyen Van Hieu, Ampy Paulo Roxas, Caesar Joventino Tado, Elmer Bautista, and Martin Gummert

Abstract The introduction of combine harvesters has made rice straw collection a major challenge and has brought bottlenecks to the rice straw supply chain. Due to this and the lack of knowledge on the straw's alternative uses, farmers burn the biomass in the field for ease of land preparation. This practice creates negative impacts on human health and the environment. However, as an alternative to burning, some Asian countries are developing increasing demands for rice straw for mushroom production, cattle feedstock, power generation, and building materials.

Mechanized straw collection has become necessary to increase capacity and to lower transportation costs. Baling machines can collect and compact rice straw in varying forms and densities. In the Mekong River Delta of Vietnam, adoption of rice straw balers have significantly improved rice straw management. A baler hauled by a 30-HP tractor has a collection capacity equal to five people, solving the labor shortage problem in rice straw collection. In addition, the volumetric weight of mechanically compacted straw bales is 50–100% higher than that of loose straw, which significantly reduces handling and transportation costs. High-density compaction (e.g., stationary compaction, briquetting, and pelletizing) can further increase the volumetric weight of baled straw from 400% to 700%, reducing transportation costs by more than 60%.

Mechanized rice straw collection and densification have contributed to improvement of the supply chain and resulted in sustainable management of rice straw. This chapter discusses the different technologies for rice straw collection, enumerating

C. Balingbing (✉) · N. V. Hung · A. P. Roxas · M. Gummert
International Rice Research Institute (IRRI), Los Baños, Laguna, Philippines
e-mail: c.balingbing@irri.org; hung.nguyen@irri.org; a.roxas@irri.org; m.gummert@irri.org

N. T. Nghi
Nong Lam University, Ho Chi Minh City, Vietnam

N. Van Hieu
Tien Giang University, Tiền Giang, Vietnam

C. J. Tado · E. Bautista
Philippine Rice Research Institute (PhilRice), Los Baños, Philippines
e-mail: cjm.tado@philrice.gov.ph; eg.bautista@philrice.gov.ph

M. Gummert et al. (eds.), *Sustainable Rice Straw Management*,
https://doi.org/10.1007/978-3-030-32373-8_2

the benefits and downsides, as well as options for further densification to reduce transportation and handling costs. The benefits and costs of various alternatives for mechanized straw collection and densification are compared and further elaborated.

Keywords Mechanized straw collection · Rice straw balers · Densification · Straw compaction · Briquetting · Pelletizing

2.1 Introduction

The intensification of rice production and rising labor costs have led to the spread of combine harvesters in Asian rice fields at harvest time. Combine harvesters leave loose rice straw on the ground, making its collection and transportation difficult, laborious, and costly. Annually, about from 600 to 800 million tons of rice straw are produced in Asia; globally approximately 1 billion tons are produced (Sarkar and Aikat 2013; McLaughlin et al. 2016). Farmers choose the quick solution of burning rice straw to quickly remove the biomass and prepare the field for the next crop. In-field burning of rice straw contributes to the emission of greenhouse gas (GHG) and poses health and environmental hazards. In addition, the potential energy that can be derived from the biomass is lost (Tabil et al. 2011).

Loose rice straw is low in density, irregular in size and shape, and difficult to handle manually. Transportation and storage of rice straw in its original form are labor-intensive and costly. The amount of rice straw available for alternative uses would be limited if there is no better way to collect it after harvest. Collecting machines make it feasible to remove a huge amount of straw in a short time (between two cropping seasons): thus, they are more economical and efficient than manual collection.

Collection of rice straw in the field using balers is becoming common in many Asian countries such as China, India, Cambodia, Vietnam, the Philippines, and Thailand, partly due to environmental regulations against field burning due to its many harmful effects (see Chaps. 8, 9, and 10). Straw needs to be gathered from the field and compressed into bales to make it compact and easy to transport. Collecting dry rice straw (moisture content at 22–32% wet basis) during the dry season is easy with a baling machine because it is lighter and does not clog the machine during baling. On the other hand, working on a wet field is quite difficult and compressing wet straw is a big challenge for the baling mechanism and requires more energy.

High-density compaction of rice straw can produce high-end market products such as high-density square bales, briquettes, and pellets, the use of which can reduce handling and transportation costs and improve processing efficiency. (Adapa et al. 2011; Emami et al. 2014).

The densification of loose biomass, such as rice straw, provides several advantages such as (1) improved handling and conveyance efficiencies throughout the supply system and biorefinery in feed, (2) controlled particle size distribution for improved feedstock uniformity and density, (3) fractional structural components for improved compositional quality, and (4) conformance to predetermined conversion technology and supply system specifications (Tumuluru et al. 2010). The common methods used to achieve densification of loose biomass, such as rice straw, includes extrusion, compacting, briquetting, or pelletizing (Demirbas and Sahin-Demirbas 2009; Tumuluru et al. 2010).

2.2 Mechanized Collection of Rice Straw

Traditionally, rice straw left after harvesting is collected by hand or with tools such as a rake or makeshift stick and carried on a canvass (Fig. 2.1), in sacks or in a carrying mat to the areas where it will be used. This method is laborious and it takes a lot of time to finish collecting all of the straw scattered in a newly harvested area. With mechanized options such as a baler (Fig. 2.2), the process is more efficient, needing only one or two skilled people to operate the machine in the field to gather the loose straw.

Fig. 2.1 Manual method of collecting rice straw after harvest in Myanmar entails considerable manual labor

Fig. 2.2 Mechanized straw collection with a small tractor drawn round baler in Vietnam

2.3 Overview of Mechanized Straw Collection Technologies

Mechanized collection of straw scattered in the field involves three main operations: (1) picking up the straw from the field, (2) compressing it into bales, and (3) transporting the bales to the bunds. In some areas, there are also some machines that just pick up the straw in loose form and transport it to the side of the field for further densification and transport.

Rice straw balers can be classified according to their mobility, the technical operation of the compacting unit, the manner of straw collection in the field, and/or the bulk density of produced bales (Fig. 2.3). A mobile baler moves on the field to collect and compress straw into bales; it can be self-propelled or pulled by a tractor. A stationary baler, on the other hand, can be used to compress rice straw disposed by stationary threshers, which are still quite common in Asia.

Balers can also be classified according to compression density (high or low), shape of bales produced (round or square), and scale (large, at least with a 100-HP tractor or engine or small, with a less than a 60-HP engine). Figure 2.4 shows a schematic diagram of a round baler gathering rice straw, forming it into cylindrical bales, and expelling them onto the field. It is pulled by a tractor connected via three points with hydraulic ports to control pick-up height. It has a series of rollers that form the round bales. The pressure for baling is delivered through the tractor's power take-off (PTO) and the bale expeller consists of a built-in, independent hydraulic mechanism. It cannot work continuously as the operation must stop periodically so that the bales can be tied with a rope and then unloaded.

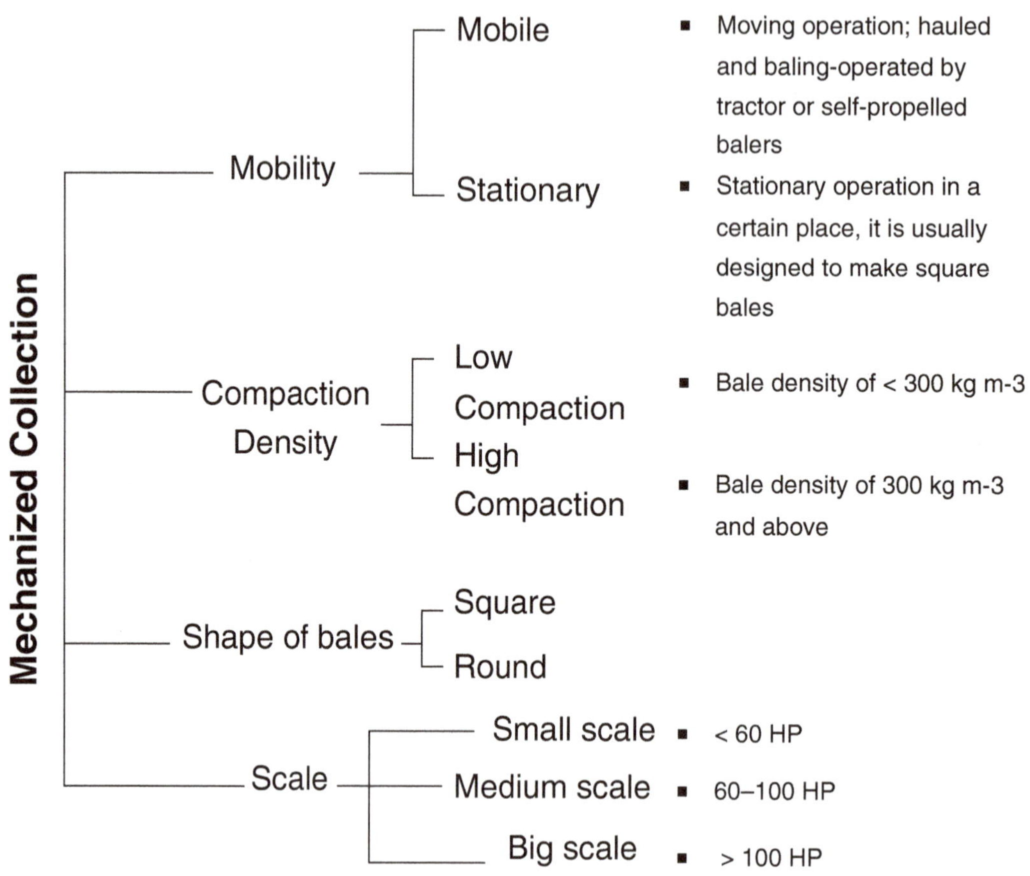

Fig. 2.3 Classification of mechanized rice straw collection in Asia

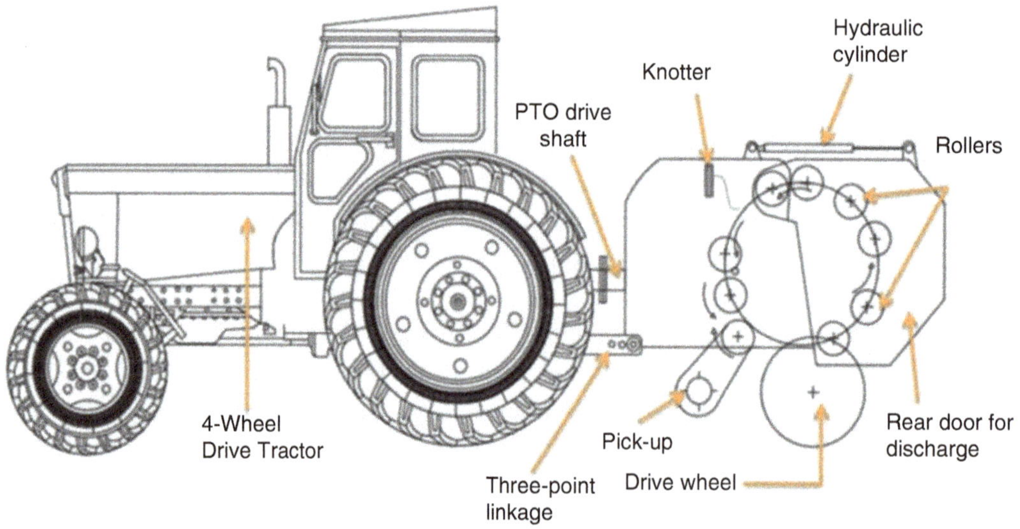

Fig. 2.4 Parts of a roller-type baler

Fig. 2.5 Square baler operating in the Philippines

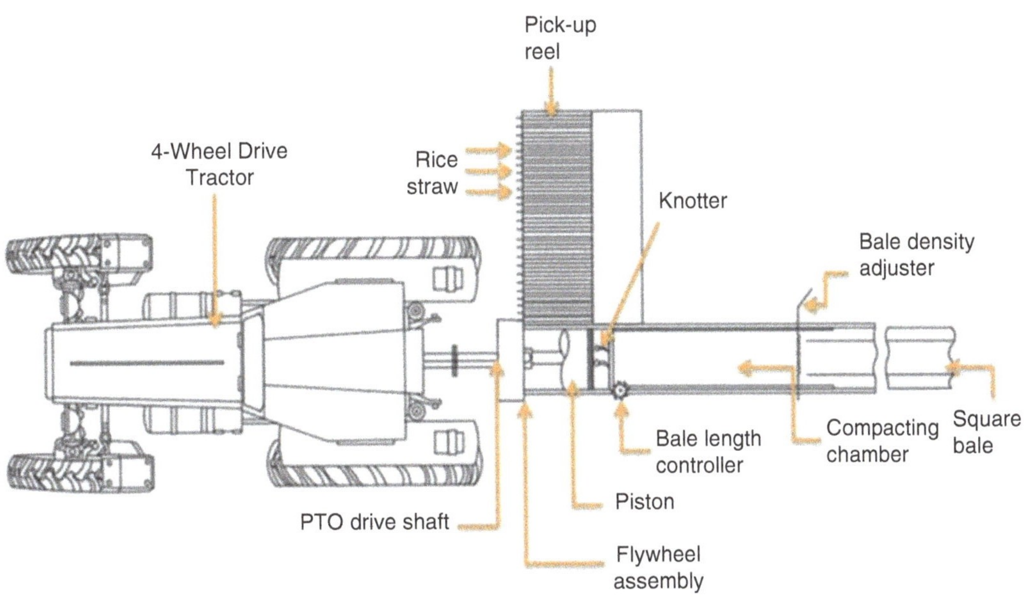

Fig. 2.6 Schematic diagram of a square baler (top view)

Figures 2.5 and 2.6 show a square baler, which uses a piston mechanism to compress loose straw into square cubes. This type of baler can operate continuously in the field without needing to stop when bales are unloaded. The baler is connected to the tractor via a drawbar hitch with power for pickup and baling delivered through the PTO. The main components of this type of baler are: (1) a pick-up reel to collect scattered straw in the field, (2) a piston to compress the straw to a set density, (3) a knotter consisting of a needle and tying mechanism, (4) a bale-length controller, and (5) s bale-density adjuster. A flywheel minimizes load peaks on the PTO generated by the reciprocal piston operation. In most models, a slip clutch on the flywheel protects the baler drive and packing system from overloading.

2.4 Commonly Used Rice Straw Balers in Asia

Small balers are better adopted and adapted in Asia as most rice fields are small at an average of about 0.05–0.4 ha (Gummert et al. 2019). Both round and square balers are adopted depending on many factors, such as soil and field conditions; preferences on bale weight; handling, transportation, storage, and multiple use purposes; and available tractors. For example, in Vietnam, farmers prefer small round balers because of their suitability for available tractors, speed in small fields, and the weight of the bales (12 kg $bale^{-1}$) produced is suitable for manual handling.

Self-propelled round balers were developed by some local manufacturers in Vietnam (Fig. 2.7a). The design is basically a round baler placed on a self-propelled undercarriage adapted from combine harvesters. It bales the straw, temporarily containing the bales on its carrier platform, and transports and discharges the bales onto the bunds. It requires a higher engine capacity compared to the tractor needed to pull the small round baler. Collection capacity is slightly lower as it moves on rubber tracks, which allow it to be used on wet fields. This and the machine's ability to move the bales to bunds have contributed to its wide adoption in the country.

Another type of self-propelled loose straw collection machine (Fig. 2.7b) was also developed based on the principle similar to the self-propelled baler except that it does not have a compacting component. This machine is used to gather scattered straw on the field and transport it loose to the side of the field.

Table 2.1 presents the features of some typical balers that are commonly used in Asia. The loose straw collection machine is normally used in a dry field for continuous operation, which is interrupted only when gathered straw is brought to the side of the field. A stationary baler can also run continuously and is typically used in a

(a) (b)

Fig. 2.7 A self-propelled baler (**a**) and a loose straw collecting machine (**b**) operating in the field in Vietnam

Table 2.1 Typical straw balers used in Asia, working characteristics, and associated costs of collection

Collection machines	Examples of manufacturer	Types of movement	Working conditions and straw location	Weight of the bales at 14% MC (kg $bale^{-1}$)	Engine power (HP)	Fuel consumption (L t^{-1})	Collection cost (US$ t^{-1})[a]
Loose-straw collection machine with rubber tracks	Phan Tan	Powered by its own engine and transmission system, typically on rubber tracks	Both dry and wet fields; equipped with a loading platform for hauling loose straw to the side of the field	NA	50	3–4	12–15
Stationary baler	Locally fabricated	Hauled to baling location	Manual feeding of straw for stationary baling; needs 3–5 operators; bale dimension is 1.5 m wide × 2.5 m long	15–20	11–25	1–2	N/A
Self-propelled baler with rubber tracks	Phan Tan	Powered by its own engine and transmission system	Both dry and wet fields; equipped with a loading platform for hauling round straw bales to the side of the field.	13–15	45	3–4	16.4
Round baler	CLAAS; STAR; John Deere	Hauled by a 4WD tractor	Operates with rollers to form round bales that are left in the field	13–15 (small); 500–600 (big)	30–80	2–3 (small); 3–4 (big)	11.3–15.4 (small)
Square baler	CLAAS; New Holland	Hauled by a 4WD tractor	Uses piston to make bales and can move continuously without stopping for unloading bales; can bale 1.5–2 t h^{-1}	15–20	50–60	3–4	12–15

Adapted from Nguyen-V-Hung et al. (2017)

[a]Collection cost includes moving straw to the side of the field, US$ t^{-1}

single location where loose straw is piled, e.g., it can be hauled, either by a two-wheel tractor or a pick-up truck, to where the stationary baler is located.

Round balers need to stop intermittently whenever bales are being discharged from the machines. Square balers, on the other hand, can be operated continuously in the field. Square bales are easy to pile and require much less space for storage than round bales. However, energy efficiency almost works out the same for both balers because a round baler has fewer power requirements than a square baler, which needs more power for compressing and baling. Round balers can also run much faster than square balers.

Rice Straw Collection in Thailand, China, and India

In Thailand, government prohibition of rice straw burning in fields has prompted farmers and the private sector to collect straw left in the field and sell it for alternative uses such as mulching and animal feedstock. The use of square balers to optimize the collection of the straw for biomass power generation has also become popular in Thailand. The cost of collection with square balers varies from US$ 18–20 t^{-1} for both of small sand large square bales in Thailand (Delivand et al. 2011).

In China, the need for systems to collect, process, and transport rice straw encouraged the introduction of many types of balers. Small, round steel-roll balers are popular in the countryside, given their simple structure and low power requirements of about 13–20 kW (Wang et al. 2011).

In India, around 120,000 t of rice straw are collected annually to add 12 megawatts of electricity to the local power grid. The huge demand for rice straw requires larger balers, such as the widely used commercial CLAAS Markant 55 (Hegazy and Sandro 2016).

2.5 High-Density Straw Compacting, Briquetting, and Pelletizing

2.5.1 Compacting

Transporting bales after collection from the field has become feasible and costs less than transporting loose straw. However, for high-end markets, such as industrial cattle farms, large amounts of rice straw (e.g., more than 20,000 t for a cattle farm in Vietnam) must be transported long distances (sometimes more than 500 km) and stored for from 3 to 6 months. Round bales should be compacted into larger and higher-density square bales to reduce transportation and storage costs.

Compacting the bales utilizes technologies that apply high pressure, such as screw or piston presses. A few compacting machines that use the piston press are found in Asia. Two common variations are the vertical and horizontal compacting systems.

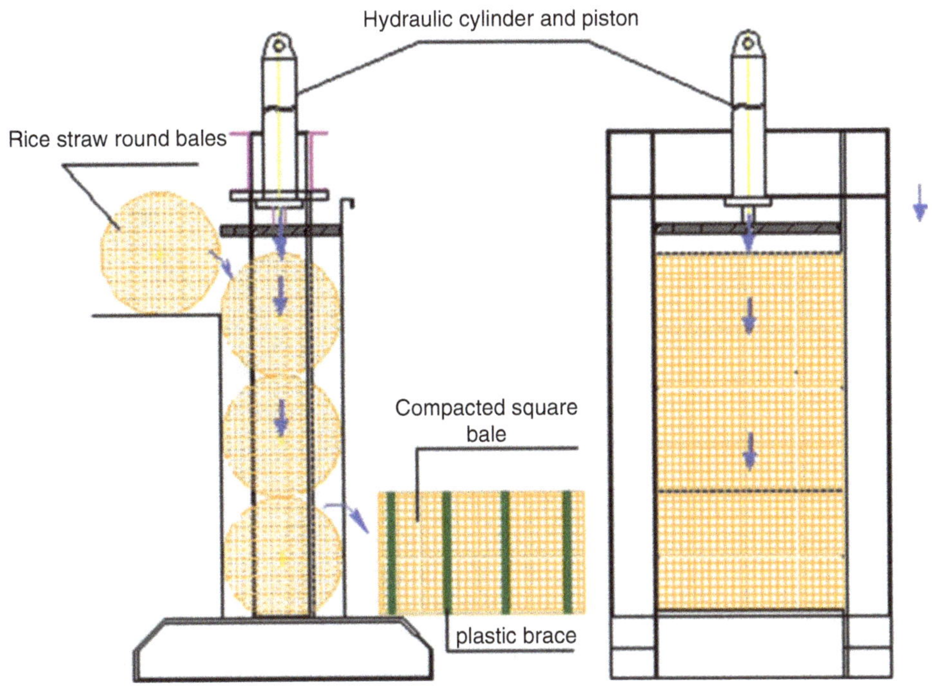

Fig. 2.8 Schematic diagram of a vertical bale compacting system

2.5.1.1 Vertical Compacting

Figure 2.8 shows a vertical compacting machine. The compacting process starts with loading straw bales into the compacting chamber through a belt conveyor. The piston vertically presses down on the bales in the chamber and then retracts before new bales are fed in. For each stroke, the piston presses three or four bales. At the end, the unloading chamber is opened to manually tie the compacted baled straws using nylon thread. After compaction, the square bales are unloaded using a forklift as shown in Fig. 2.9.

2.5.1.2 Horizontal Compacting

A schematic diagram of a horizontal compacting machine is shown in Fig. 2.10. Its mechanism is basically the same as the vertical system. The only difference is that it has a horizontal compressing direction. The main advantage is that the operation can be automated through conveyor belts instead of having an operator that is needed for the vertical compacting machine for loading and unloading. Additional advantages are consistent density of compacted materials and higher volume compared to vertical compactors (Fig. 2.11). On the other hand, vertical compactors use far less space and cost much less than horizontal ones.

Fig. 2.9 A vertical bale compacting system in operation

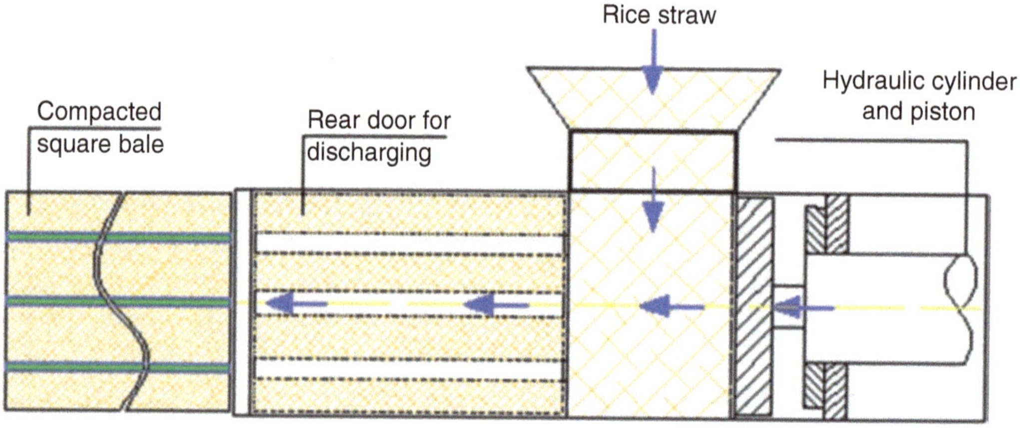

Fig. 2.10 Schematic diagram of horizontal compressing system

2.5.2 *Briquetting*

Briquettes (Fig. 2.12) are produced by compressing chopped straw into a cylindrical form through a briquetting press shown in Fig. 2.13. The hydraulic operation pressure of the briquetting press (Muetek MPP 130) is set to 15 MPa and the press works in three stages. First, the feedstock passes through the pre-compression unit, which presses it inside the pressing block in a Y-direction. When a resistance of

Fig. 2.11 Horizontal compressing system

Fig. 2.12 Rice straw briquettes

8 MPa is achieved, the piston starts to operate and densifies the feedstock in an X-direction. At the open end of the pressing block, a pressing clamp is installed that opens to eject the produced briquette. The pressure at which the pressing clamp opens, which varies from 4 to 8 MPa, is related to the compaction density of the briquettes (Munder 2013).

Briquettes are also produced through: (1) the press-chamber principle, which consists of two parts: a heated die that acts as a press and a punch that fits in tight; (2) the screw principle, based on continuous extrusion of the feedstock by a screw through a heated tape die; and (3) a piston press, where a reciprocating ram presses the straw biomass in a die. The finished products would have varying energy density depending on the technology used (Munder 2013).

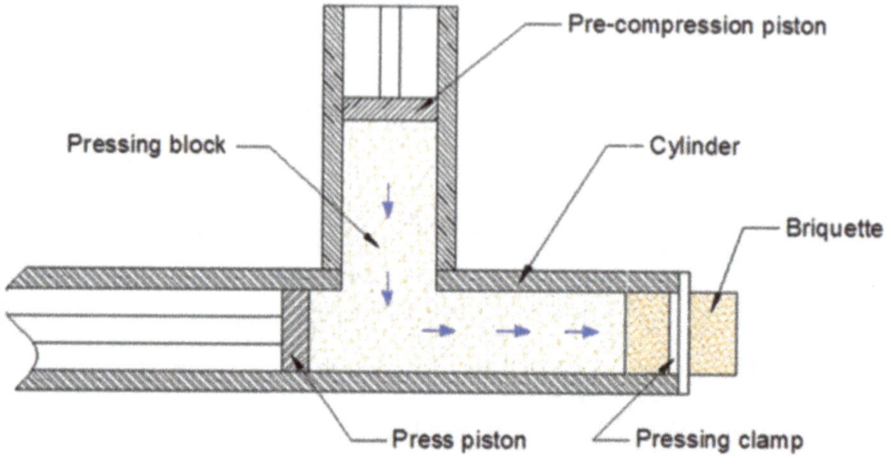

Fig. 2.13 Schematic diagram of the pressing unit

Fig. 2.14 Rice straw pellets

As fuel briquettes have an advantage over loose rice straw in terms of higher volumetric calorific value, improved combustion characteristics, ease of use when feeding the furnace, and uniformity in size and shape. A rice straw briquette has an average length of 10 mm (Munder 2013) and a density of up to 0.97 g cm^{-3}, which is 48 times the density of loose rice straw.

2.5.3 Pelletizing

Pellets (Fig. 2.14) are produced based on the principle of compressing ground straw. As shown in Fig. 2.15, the compression unit is composed of a horizontal or vertical ring die and rollers that put pressure on the inner surface of the ring die. During the pelletizing process, the ring die and rollers rotate and the raw materials fall in the

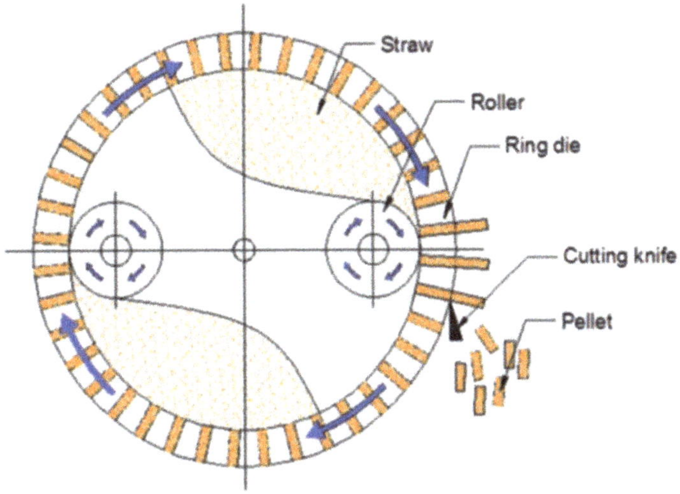

Fig. 2.15 Schematic diagram of ring die pelletizing Biomass pelletizing usually involves chopping, grinding, mixing, pelletizing, and packaging with the corresponding components shown in Figs. 2.16 and 2.17. Straw is chopped using a rotating chopper. Then, it is fed into a grinding machine. After passing through the grinding machine, the straw is collected using a cyclone and is temporary stored in a silo. Finally, the ground straw is loaded into a pelletizing machine. Pellet quality is influenced by the characteristics of the feeding materials and operating conditions of the pelletizing process both of which are controllable (Rhen et al. 2005; Tumuluru 2014). Basic parameters that are necessary in the pressing process include a pressing pressure of 80 MPa and a temperature of 105 °C

clearance between the ring die and the roller, which are pressed into the holes on the ring die. Pellets are cut at the outer surface of the ring die and collected.

Compared with other compacting processes, such as briquetting and tumble agglomeration, pellets are generally regarded as the most durable because they are placed under the highest amount of pressure during formation (Whittaker and Shield 2017). Pelletizing can increase the bulk density of the biomass from an initial value of 40–200 kg m^{-3} to a final bulk density of 600–800 kg m^{-3}. Pelletizing can overcome hurdles in cost and logistics in utilizing loose straw for energy or animal feedstock.

The product quality and calorific value of straw to be pelletized can be improved by mixing it with various additives, such as starch, molasses, ash, montan resin, paraffin, palmitin, and anthracite and lignite coal. The compressing pressure is the most significant factor affecting pellet density and the biomass type significantly affects pellet durability (Adapa et al. 2011). The physical quality of compacted loose biomass materials is partly indicated by compressive strengths, durability, stability, and smoothness (Demirbas and Sahin-Demirbas 2009). The specific energy requirements of different types of biomass for compression vary according to the compressed density of materials and the moisture content of biomass inputs. Density is identified as an important parameter in compression, i.e., the higher the density, the higher the energy/volume ratio.

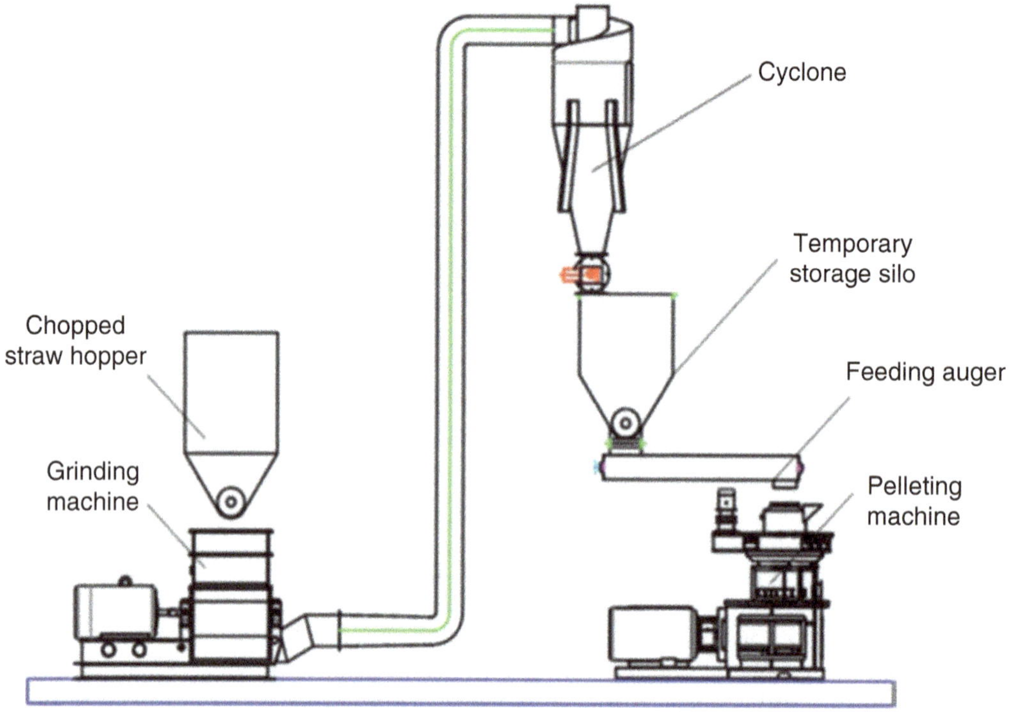

Fig. 2.16 Schematic diagram of pelletizing system. (Adapted from Nguyen-Van-Hieu et al. 2018)

Fig. 2.17 A rice husk pelletizing machine. (Adapted for pelletizing rice straw)

Pelletized rice straw can be used as fuel, animal feedstock, or material for anaerobic digestion. The pelletizer die hole size is known to have an important effect on the moisture content of the pellet, while the temperature reached during pelletization can also influence pellet quality.

Straw densification through pelletizing can increase bulk density from 600 to 800 kg m^{-3} (Kaliyan and Morey 2009; Kargbo et al. 2009). The average specific mass of a straw pellet may also reach 1244 kg m^{-3}, which is higher by 1000% compared with loose straw. Said et al. (2015) reported that the ideal value for high-quality pellets is 1200 kg m^{-3}. Pelletized rice straw has an advantage of preventing straw materials from floating in water when using the straw for other processes such as anaerobic digestion. The use of enriched pellets as feed for cows results in minimal waste and leftovers during feeding.

The production costs of rice straw pellets are computed based on the estimated cost of equipment and assumed cost of straw and labor at the locality (including depreciation, material, and labor costs). In one case study in Vietnam (Nguyen-Van-Hieu et al. 2018), materials (straw and cattle feed additives) cost US$ 280 t^{-1}; straw prices ranged from US$ 90 to 100 t^{-1}; and depreciation, labor, and electricity costs were estimated based on the existing rice husk pelletizing system. Given a pelletizing cost of US$ 22.6 t^{-1}, straw pellets cost approximately US$ 125 t^{-1}. Pelletizing can significantly reduce transportation costs. In the same case study, a cubic bale was sold at a price of US$ 110 t^{-1} excluding transportation cost, which was about US$ 35.5 t^{-1} for a distance of 1000 km by truck. The cost of grinding straw was estimated at US$ 100 t^{-1}. Transporting pelletized straw was found to be more economical and practical compared to bales.

2.6 Conclusions and Recommendations

Alongside the spread of combine harvesters, government regulations against open field burning of rice straw, and increasing use of straw, mechanized collection is gaining ground in Asia. Small balers with a capacity of 1–2 t h^{-1}, which are easy to maneuver in small fields, have been found suitable in Cambodia, the Philippines, and Vietnam. The self-propelled baler—a successful innovation in Vietnam—is being adapted in Southeast Asian countries, such as Cambodia and the Philippines, because it reduces labor requirements in hauling baled straw from the field to the bund. Another advantage is the machine's rubber chain-wheel mechanism, which makes it suitable for use in wet fields, particularly in areas where field drainage is a problem.

A case study in Vietnam showed that mechanized collection can reduce costs by about 68% compared to manual collection. As labor scarcity rises, machines become a more sustainable option for Asian rice fields where farmers have traditionally resorted burning straw after harvest, which is easier and cheaper.

As Asian countries move towards field consolidation and upgrading of contractual arrangements among farmers, mechanized collection is likely to become more efficient. Further research has to be conducted to understand field efficiency vis-à-vis field capacity so that (1) the use of baler machines is optimized, (2) the sustainability of custom servicing business models is assured; and (3) machine owners are adequately informed on the viability of their investments.

Rice straw densification—through compacting, briquetting, or pelletizing—results in better handling and storage of the byproduct, which, in turn, reduces transportation costs and makes efficient use of storage facilities. The technologies now available, such as briquette presses and pelletizers, also provide options for other uses of rice straw, such as animal feed, fuel, and feedstock for energy generation.

The processing of loose straw into pellets can further save transportation costs and improve logistical processes as experienced in Vietnam. Research is still required to improve the quality of densified straw, either for animal feed or fuel. Researchers should look into locally available binding materials that are cheap and of high quality to improve pellet and briquette properties in terms of strength, durability, density, nutrition (for animal feed), and calorific value (for fuel).

References

Adapa P, Tabil L, Schoenau G (2011) A comprehensive analysis of the factors affecting densification of barley, canola, oat and wheat straw grinds. Written for presentation at the CSBE/SCGAB 2011 Annual Conference Inn at the Forks, Winnipeg, Manitoba. 10–13 July 2011

Delivand MK, Barz M, Gheewala SH (2011) Logistics cost analysis of rice straw for biomass power generation in Thailand. Energy 36:1435–1441. https://doi.org/10.1016/j.energy.2011.01.026

Demirbas K, Sahin-Demirbas A (2009) Compacting of biomass for energy densification. Energ Source Part A: Recov Utilization Environ Effects 31(12):1063–1068. https://doi.org/10.1080/15567030801909763

Emami S, Tabil GL, Adapa P, George E, Tilay A, Dalai A, Drisdelle M, Ketabi L (2014) Effect of fuel additives on agricultural straw pellet quality. Int J Agric Biol Eng 2:92–100

Gummert M, Quilty J, Hung NV, Vial L (2019) Engineering and management of rice harvesting. In: Pan Z, Khir R (eds) Advances in science and engineering of Rice. DEStech Publications, Inc., Lancaster, pp 67–102

Hegazy R, Sandro J (2016) Report: rice straw collection. Available at https://www.researchgate.net/publication/301770258

Kaliyan N, Morey RV (2009) Factors affecting strength and durability of densified biomass products. J Biomass Bioenerg 33:337–359

Kargbo FR, Xing J, Zhang Y (2009) Pretreatment for energy use of rice straw: a review. Afr J Agric Res 4(13):1560–1565

McLaughlin O, Mawhood B, Jamieson C, Slade R (2016) Rice straw for bioenergy: The effectiveness of policymaking and implementation in Asia. https://www.researchgate.net/publication/305882098

Munder S (2013) Improving thermal conversion properties of rice straw by briquetting. Masters thesis, Nachwachsende Rohstoffe und Bioenergie. Unbiversitat Hohenheim, Institute Fur Agrartechnik

Nguyen-Van-Hieu, Nguyen-Thanh-Nghi, Le-Quang-Vinh, Le-Minh-Anh, Nguyen-Van-Hung, Gummert M (2018) Developing densified products to reduce transportation costs and improve the quality of rice straw feedstocks for cattle feeding. J Vietnamese Environ 10(1):11–15

Nguyen-V-Hung, Balingbing C, Quilty J, Sander BO, Demont M, Gummert M (2017) Processing rice straw and rice husk as co-products. In: Sasaki T (ed) Achieving sustainable cultivation of rice, vol 2. Burleigh Dodds Science Publishing, Cambridge, pp 121–148

Rhen C, Gref R, Sjostrom M, Wasterlund I (2005) Effects of raw material moisture content, densification pressure and temperature on some properties of Norway spruce pellets. Fuel Process Technol 87(1):11–16. https://doi.org/10.1016/j.fuproc.2005.03.003

Said N, Abdel daiem MM, Garcia-Maraver A, Zamorano M (2015) Influence of densification parameters on quality properties of rice straw pellets. Fuel Process Technol 138:56–64

Sarkar N, Aikat K (2013) Kinetic study of acid hydrolysis of rice straw. Hindawi Publishing Corporation. ISRN Biotechnology https://doi.org/10.5402/2013/170615

Tabil L, Adapa P, Kashaninejad M (2011) Biomass feedstock preprocessing–Part 2: Densification. In Bernardes MADS (ed) Biofuel's Engineering Process Technology, p. 439–464

Tumuluru JS (2014) Effect of process variables on the density and durability of the pellets made from high moisture corn Stover. Biosyst Eng 119:44–57

Tumuluru JS, Wright CT, Kenny KL, Hess JR (2010) A review on biomass densification technologies for energy application. Idaho National Laboratory, Biofuels and Renewable Energies Tehcnologies Department, Energy Systems and Technologies Division, Idaho Falls, Idaho 83415. Available at https://pdfs.semanticscholar.org/305d/4fb8d8c782e40caa96847d555e02159b9c74.pdf

Wang D, Buckmster DR, Jiang Y, Hua J (2011) Experimental study on baling rice straw silage. Int J Agric & Biol Eng 1. https://doi.org/10.3965/j.issn.1934-6344.2011.01.000-000

Whittaker C, Shield I (2017) Factors affecting wood, energy grass and straw pellet durability. A review. Renew Sust Energ Rev 71:1–11

Chapter 3
Rice Straw-Based Composting

Nguyen Thanh Nghi, Ryan R. Romasanta, Nguyen Van Hieu, Le Quang Vinh, Nguyen Xuan Du, Nguyen Vo Chau Ngan, Pauline Chivenge, and Nguyen Van Hung

Abstract Current practices in rice production leave a huge amount of wet straw on the field, which cannot be used as feed or for food production. Compost production is one way of effectively utilizing rice straw. Spent rice straw from mushroom production is also used as compost but this has low nutrient value and is poorly decomposed when using it as a soil improver. This wet, low-quality straw, as well as byproducts from mushroom and cattle feed production, could be used to produce better-quality compost to return nutrients back to the field. Mechanization in mixing the materials, i.e., a compost turner, is necessary to have good aeration, increase the decomposition process, and reduce labor cost. This chapter provides an overview of composting technology and current practices of rice-straw composting. Updated information on this topic, resulting under the current BMZ-funded IRRI rice-straw management project (2016–2019), which has been implemented in Vietnam and the Philippines, is also included here, particularly in the sections on vermin-composting and mechanized composting.

Keywords Rice straw · Compost turner · Organic fertilizer

N. T. Nghi (✉) · L. Q. Vinh
Nong Lam University, Ho Chi Minh City, Vietnam

R. R. Romasanta · P. Chivenge · N. V. Hung
International Rice Research Institute (IRRI), Los Baños, Laguna, Philippines
e-mail: r.romasanta@irri.org; p.chivenge@irri.org; hung.nguyen@irri.org

N. Van Hieu
Tien Giang University, Tien Giang, Vietnam

N. X. Du
Sai Gon University, Ho Chi Minh City, Vietnam

N. V. C. Ngan
Can Tho University, Can Tho, Vietnam
e-mail: nvcngan@ctu.edu.vn

M. Gummert et al. (eds.), *Sustainable Rice Straw Management*,
https://doi.org/10.1007/978-3-030-32373-8_3

3.1 Overview of Composting Technology

Composting converts organic mass, such as rice straw, other agricultural by-products, digestive, animal wastes, etc., into a more decomposed product, called compost. Composting is necessary because it can help to increase significantly the quality of the compost product based on the optimized nutrient factors and the decomposition process (Diaz et al. 2007). Compost is used as a soil improver or directly as a planting substrate. Application of compost results in an increase in, not only crop yield, but also soil fertility (Goyal et al. 2009; Vo-Van-Binh et al. 2014).

Compost quality is strongly affected by the factors happening during the composting process, such as temperature, pH, carbon-to-nitrogen ratio (C/N), etc. These factors can be controlled through bio-, chemical-, and physical methods or a combination of these to optimize the composting processes and products. This chapter provides an overview of the main factors including temperature, pH, C/N ratio, moisture content, and properties of the feedstock.

3.1.1 Properties of Materials

In composting, waste products are mixed followed by creating conditions to enable biological decomposition to result in a higher-quality organic resource. The speed of the composting process and the quality of compost depend on the type, quality, and chemical and physical properties of the raw materials, conditions, and environment during the process. Typical physical and chemical properties of different raw materials for composting are shown in Table 3.1.

3.1.2 Temperature

Temperature effects on the composting process can be divided into the four phases, (1) mesophilic, (2) thermophilic, (3) cooling, and (4) maturing. During the initial phase of decomposition and break down of compounds, heat is generated due to the bio-oxidative microbial degradation (Diaz et al. 2007). This phase is facilitated by mesophilic bacteria, which become less competitive as temperature increases up to approximately 40 °C when thermophilic bacteria become predominant. At about 55 °C, destruction of plant pathogens occurs (Shilev et al. 2007) and then complete hygienization takes place at temperatures of 60 °C and above (Shilev et al. 2007). However, temperatures exceeding 65 °C should be avoided as they may harm even useful microbes (Shilev et al. 2007). According to Haug (1980), the composting temperature has to be above 55 °C for three consecutive days to kill the pathogens. The temperature of compost reaches 60 °C after 10 days and lowers to ambient temperature from 60 to 90 days of composting (Jusoh et al. 2013). Temperature

Table 3.1 Chemical and physical properties of raw materials for composting

Feedstock	Properties						Sources
	pH	TOC (%)	C (%)	N (%)	C/N ratio	MC (%)	
Rice straw	7.6	39.2			61.3	11.4	Jusoh et al. (2013)
			40.2	0.7	55.1	10.2	Qiu et al. (2013)
			47.0	1.3	35.3		IRRI (2019)
Spent rice straw after mushroom production			14.3	0.7	21.9		
			13.3	0.9	14.3		
Banana trunk					39.6	90	
Sawdust			50.8	0.8	60.4	4.6	Qiu et al. (2013)
Green waste	6.5	15.3			8.4	79.0	Jusoh et al. (2013)
Goat manure	7.1	35.6			13.0	58.0	
Cow manure			12.9	0.9	14.5		IRRI (2019)
			11.4	0.8	14.0		
Hog manure			15.3	0.9	16.5	53.6	Qiu et al. (2013)

TOC total organic carbon, *C/N ratio* carbon/nitrogen ratio, *MC* moisture content, *N* nitrogen, *P* phosphorus, *K* Potassium

could be controlled effectively through aeration and mixing in the cooling-down stage. The final humus is then produced after the maturing phase.

Temperature patterns during the composting process with effective microorganisms (EMs) and without EMs similarly have been shown to fluctuate over time (IRRI 2019; Jusoh et al. 2013). These studies conducted the experiments on vermicomposting using rice straw, cow manure, shredded banana trunk, and African Night Crawler (ANC). The temperature profile during composting is shown in Fig. 3.1. Temperature of the composts reaches to 70 °C during the first and second phases (thermophilic) of the composting process, and then cools down to from 22 to 35 °C finally balancing with the ambient temperature.

3.1.3 pH Value of Composting Environment

The pH also strongly affects the composting process. High pH together with high temperature at the beginning of composting can cause a loss of nitrogen through ammonia volatilization (Diaz et al. 2007). Generally, pH of organic matters used for composting widely varies from 3 to 11 (Bertoldi et al. 1984). In some specific research, the optimized pH value is 7.60 ± 0.08 for rice straw, 7.10 ± 0.08 for goat manure, and 6.5 ± 0.48 for green waste (Jusoh et al. 2013). In research results under

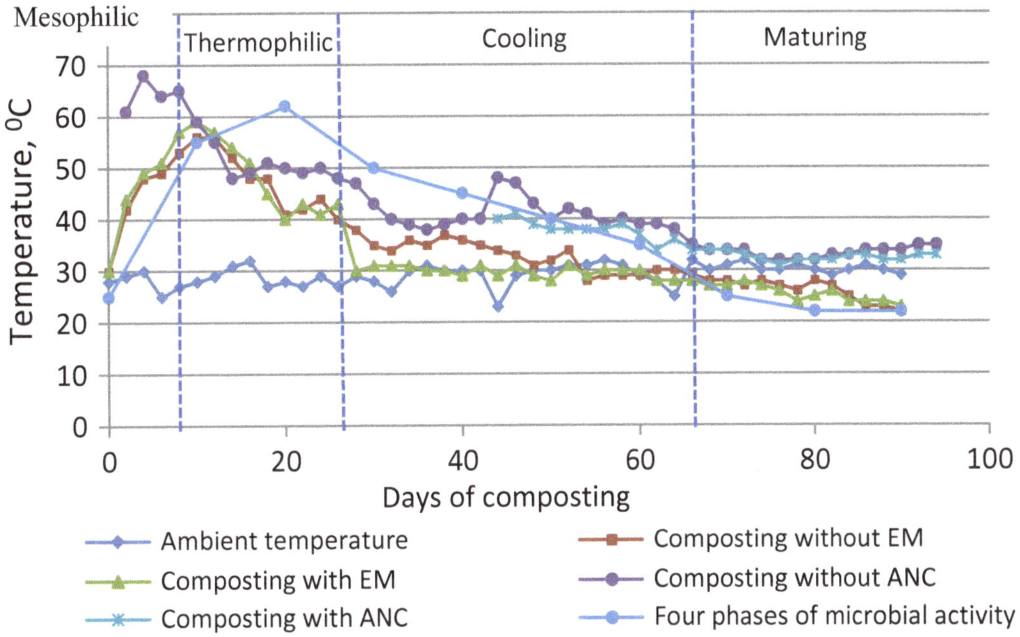

Fig. 3.1 Temperature profile during composting of rice straw mixed with cow manure, shredded banana trunk, and ANC, adapted from Jusoh et al. (2013), Shilev et al. (2007), and IRRI (2019)

the BMZ-IRRI rice-straw management project (IRRI 2019), the pH value of compost after production was from 6.80 to 6.85 (Table 3.1), which are in line with the recommended range of from 6.9 to 8.3 at the end of composting (Ameen et al. 2016; Diaz et al. 2007; El-Haddad et al. 2014).

3.1.4 *Carbon-to-Nitrogen Ratio*

The ratio of carbon-to-nitrogen (C/N ratio) in the mixing compound depends on the C/N ratio of the materials and its mixing ratio. It is computed based on the Eq. (3.1):

$$C/N = \frac{W_1 * C_1 + W_2 * C_2 + \ldots + W_n * C_n}{W_1 * N_1 + W_2 * N_2 + \ldots + W_n * N_n} \tag{3.1}$$

Where:

$W_1, W_2, \ldots W_n$ = weight of single materials
$C_1, C_2, \ldots C_n$ = organic carbon content of single materials
$N_1, N_2, \ldots N_n$ = Nitrogen content of single materials

The C/N ratio of the residue to be composted is one the most important factors affecting the quality and period of composting. In addition, the composition and the mixing ratio of the raw materials used for composting influence the quality of the compost. For an optimal process, a C/N ratio in the range of 20–30 is generally recommended (Vigneswaran et al. 2016). Higher ratios lead to longer composting periods due to slower decomposition whereas, at lower ratios, nitrogen can be rapidly lost by conversion into gaseous forms (as ammonia or nitrogen), affecting the compost quality. Thus, it is necessary to mix straw with a high-quality organic material such as manure in order to have a suitable ratio of C/N (25–30) for composting process (Ameen et al. 2016). Due to the low C/N ratio of cattle manure, it was adjusted to 25 by adding sawdust or rice straw (Qiu et al. 2013; Jusoh et al. 2013).

3.1.5 Moisture Content of Substrates

The moisture content of the composting materials affects the availability of oxygen for microbial processes. Water is essential for the decomposition process and water stress is among the most common limitations on microbial activity on solid substrates. However, when moisture levels exceed 65%, air in the pore spaces of the raw materials is displaced by water, which leads to anaerobic conditions, odors, and slower decomposition (Pace et al. 1995; Sherman 1999). Moisture content of the mixture was maintained at 60%, which is the optimum level for microbial activity (Goyal and Sindhu 2011; Diaz et al. 2007). To maintain the moisture content at an optimal range of 50–65% (wet basic), water is added to the compost during turning periods. After the turning process, a plastic sheet is used to cover the windrow to retain the moisture content and prevent excessive loss of heat (Vigneswaran et al. 2016; Jusoh et al. 2013). At the end of the composting process, the moisture content of the compost should be about 30% to prevent any further biological activity in the stabilized material (Diaz et al. 2007).

3.2 Current Practices for Rice-Straw Composting

3.2.1 Vermi-composting

Vermi-composting is a biological process of bio-oxidation and stabilization of organic material involving the joint action of earthworms and microorganisms (Aira et al. 2002). While microbes are responsible for the biochemical degradation of organic matter, earthworms serve as the important drivers of the process in conditioning the substrate and alteration of biological activity. The end product, or vermicompost, is a finely divided peat-like material with high porosity and water-holding capacity that contains most nutrients in forms that are readily taken up by the plants (Dominguez and Edwards 2004). The invertebrates have indirect effects on the structure and activities of bacterial and fungal communities through inoculum dispersal, grazing, litter combination, gut passage, and aggregate formation (Anderson 1987). The use of earthworms in compost heaps, beds or boxes makes the process

Fig. 3.2 ANCs are incorporated in the windrows after the anaerobic stage of decomposition (left) and harvested around 80–90 days (right)

faster for the breakdown of organic waste and its decomposition, i.e., composting (Edwards et al. 1989; Gaur and Sadasivam 1993).

Vermi-composting technology is employed using windrows for composting. A windrow consists of layers rice straw, manure, and shredded banana trunks. Water is added during windrow building to reach a moisture content of 60%, which is a suitable condition for composting. A field trial of vermi-composting was conducted in 2017–2018 at IRRI in the Philippines (IRRI 2019). The experiment was set up with a windrow height of 1 m, a width of 1.5 m. It was composed of four layers of rice straw, cow manure, and shredded banana trunk (Fig. 3.2). From the total amount produced in every windrow, it is expected to recover 50% of the vermicast.

The vermi-compost consists of two composting stages, anaerobic and aerobic. Anaerobic composting is implemented during the first 40 days by covering the compost heap with a plastic sheet that reduces the exchange of air between the atmosphere and the compost. The covers are then removed for the next 40–50 days. The ANC, which is introduced during the aeration phase, is one the popular species of earthworm used for this process. Watering of the windrows is also essential for ANC to thrive, grow, and be efficient in producing vermicast. Water is applied for every windrow, 100 L for 1000 kg of composting materials for every other day from day 40 to day 80, and daily from day 81 to day 93. To efficiently manage water use, drip-irrigation technology is recommended. The vermicast recovery ratio is 1:2, which means that, with a total input of 1000 kg of compost, 500 kg of vermicast are recovered.

3.2.2 Mechanized Windrow Composting

The windrow-composting method consists of linear rows of compost materials (rice straw and cow manure), which are placed layer by layer and mechanically turned periodically. The air contained in the interspaces of the composting mass varies in composition. The CO_2 content gradually increases and the O_2 level falls during composting process. The concentration of O_2 for composting varies from 15 to 20%. (Diaz et al. 2007). Thus, the turning process helps improve aeration and mixing of compost constituents. The windrow composting method relies on mechanical aeration, typically with a compost windrow turner, to optimize the composting process. During turning, microbial inoculum is mixed with water and sprayed in the windrow to speed up the composting process and obtain the required moisture content. The interval between turnings is usually 10–14 days (IRRI 2019), but 15-day intervals have also been reported (Muzamil 2012). After turning, the windrow is covered using a plastic sheet to maintain the proper moisture content and temperature. The height and width of the windrows are typically set to fit the size of the turner.

A mechanical windrow composting system comprises main components of turner and tractor (Fig. 3.3a; IRRI 2019). A windrower or compost turner presented in IRRI (2019) comprises six main parts: turning drum, universal joints, trail linkage, wheels, gear box, and frame and housing (Fig. 3.3b).

The turner is pulled by a tractor through a trail linkage system. The rotor of the turner is powered by the tractor's power-take-off shaft. The blades installed on the drum rotate to turn the materials in the windrow when the machine moves forward. After the turning, the substrates are pushed to the middle of windrow. According to IRRI (2019), the turner hauled by a 30- to 50-HP tractor has a capacity of 30 t/h.

The resulting compost can help improve rice productivity (see Chap. 9) and other crops, particularly vegetables. Additionally, the application of the compost to rice production decreases greenhouse gas emissions compared to when fresh straw is incorporated in situ (see Chap. 10). The process serves to bring value to the waste products that would otherwise have environmental consequences.

Fig. 3.3a Mechanical windrow composting system with tractor and turner

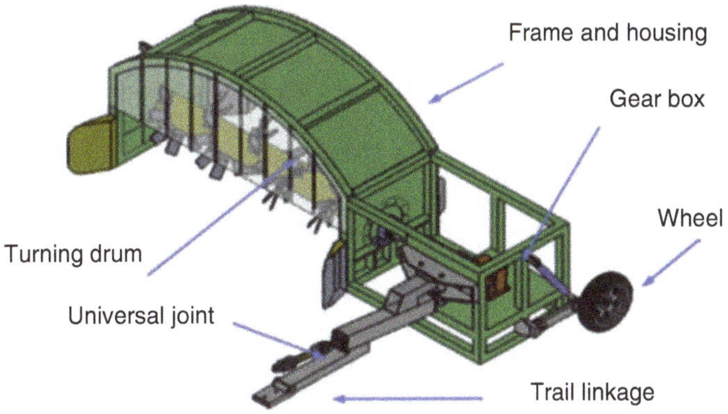

Fig. 3.3b Structure of the compost turner designed under the BMZ-IRRI project (IRRI 2019)

3.3 Conclusions and Recommendations

The best practices of composting described here can help optimize the quality and nutrient efficiency of the mixture of rice straw and animal manure used to improve soil and crop productivity.

The application of composting technology and the compost turner can contribute to reducing labor costs in turning, creating alternative uses for rice straw, and increasing farmers' income by adding value to rice and other related uses, such as mushroom production. In addition, increasing the value of the rice straw, especially low-quality straw, leads farmers to avoid burning it in the field.

References

Aira M, Monroy F, Dominguez J, Mato S (2002) How earthworm density affects microbial biomass and activity in pig manure. Eur J Soil Biol 38:7–10

Ameen A, Ahmad J, Munir N, Raza S (2016) Physical and chemical analysis of compost to check its maturity and stability. Eur J Pharm Med Res 3(5):84–87

Anderson JM (1987) Interactions between invertebrates and microorganisms: noise or necessity for soil processes. In: Anderson JM, ADM R, DWH W (eds) Ecology of microbial bommunities. Cambridge University Press, Cambridge, pp 125–145

Bertoldi MD, Vallini G, Pera A (1984) Technology aspect for composting including modeling and microbiology. In: JKR G (ed) Composting of agricultural and other wastes. Elsevier Applied Science Publisher, London, pp 27–40

Diaz LF, de Bertoldi M, Bidlingmaier W, Stentiford E (2007) Compost science and technology. Waste management series, 8, Elsevier, Boston

Dominguez J, Edwards CA (2004) Vermicomposting organic wastes: a review. Soil zoology for sustainable development in the 21st century, Cairo, pp 369–395

Edwards WM, Shipitalo MJ, Owens LB, Norton LD (1989) Water and nitrate movement in earthworm burrows within long-term no-till cornfields. J Soil Water Conserv 44(3):240–243

El-Haddad ME, Zayed MS, El-Sayed GAM, Hassanein MK, Abd El-Satar AM (2014) Evaluation of compost, vermi-compost and their teas produced from rice straw as affected by addition of different supplements. Ann Agric Sci 59(2):243–251

Gaur AC, Sadasivam KV (1993) Theory and practical considerations of composting organic wastes. Org Soil Health Crop Prod:1–21

Goyal S, Sindhu SS (2011) Composting of rice straw using different inocula and analysis of composting quality. Microbiol J. Haryana Agricultural University, India

Goyal S, Singh D, Suneja S, Kapoor KK (2009) Effect of rice straw compost on soil microbiological properties and yield of rice. Indian J Agric Res 43(4):263–268

Haug RT (1980) Compost engineering: principles and practice. Ann Arbor Science Publishers, Michigan

IRRI (International Rice Research Institute) (2019) Final report of the BMZ-funded IRRI sustainable rice straw management project (unpubl)

Jusoh ML, Manaf CLA, Latiff PA (2013) Composting of rice straw with effective microorganisms (EM) and its influence on compost quality. Iranian J Environ Health Sci Eng 10(1):17. https://doi.org/10.1186/1735-2746-10-17

Muzamil M (2012) Optimization of design and operational parameters of compost-turner-cum-mixer. Master thesis. Indian Agricultural Research Institute New Delhi

Pace MG, Miller BE, Farrell-Poe KL (1995) The composting process. Utah State University Extension

Qiu J, He J, Liu Q, Guo Z, He D, Wu G, Xu Z (2013) Effects of conditioners on sulphonamides degradation during the aerobic composting of animal manures. Procedia Environ Sci 16:17–24

Sherman R (1999) Large-scale organic materials composting. North Carolina Cooperative Extension Service https://content.ces.ncsu.edu/large-scale-organic-materials-composting

Shilev S, Naydenov M, Vancheva V, Aladjadjiyan A (2007) Composting of food and agricultural wastes. https://doi.org/10.1007/978-0-387-35766-9_15. https://www.researchgate.net/publication/226659156_Composting_of_Food_and_Agricultural_Wastes

Vigneswaran S, Kandasamy J, Johir MAH (2016) Sustainable operation of composting in solid waste management. Procedia Environ Sci 35:408–415

Vo-Van-Binh, Vo-Thi-Guong, Ho-Van-Thiet, Le-Van-Hoa (2014) Long-term effects of application of compost on soil fertility improvement and increase in yield of rambutan fruit in Ben Tre Province (in Vietnamese). Scientif J Can Tho Univ 3:133–141

Chapter 4
Thermochemical Conversion of Rice Straw

Monet Concepcion Maguyon-Detras, Maria Victoria P. Migo, Nguyen Van Hung, and Martin Gummert

Abstract Biomass conversion into various forms of energy, such as heat, power, or biofuels using thermal processes, involves the decomposition of biomass by exposure to heat, typically above 300 °C. Thermal conversion processes include pyrolysis, gasification, and direct combustion. Several factors affect the yields and energy recovery from these processes including temperature, reaction time, heating rate, absence, or presence of oxygen, use of catalysts, and pressure. Due to rice straw's relatively high carbon and hydrogen contents, it contains a considerable amount of energy that make it a suitable feedstock for thermal conversion. In this chapter, the basic principles and factors affecting the thermal conversion of biomass into energy are discussed. Studies on the use of rice straw as feedstock to produce heat, power, and biofuels via thermal conversion are reviewed. Utilization of thermal conversion byproducts including biochar and ash will are presented. Thermal processes are compared in terms of energy conversion, possible environmental impacts, and technological and commercial maturity.

Keywords Bioenergy · Pyrolysis · Gasification · Combustion · Biomass · Thermochemical conversion

---

M. C. Maguyon-Detras (✉) · M. V. P. Migo
Department of Chemical Engineering, College of Engineering and Agro-Industrial Technology, University of the Philippines Los Baños (UPLB), Los Baños, Laguna, Philippines
e-mail: mmdetras@up.edu.ph; mpmigo@up.edu.ph

N. V. Hung · M. Gummert
International Rice Research Institute (IRRI), Los Baños, Laguna, Philippines
e-mail: hung.nguyen@irri.org; m.gummert@irri.org

M. Gummert et al. (eds.), *Sustainable Rice Straw Management*,
https://doi.org/10.1007/978-3-030-32373-8_4

4.1 Overview of Thermal Conversion Processes

Biomass conversion by thermal processes (i.e., direct combustion, gasification, pyrolysis) involves the decomposition of biomass into various solid, liquid, and gaseous components by exposure to heat (normally above 300 °C). Different process conditions, such as temperature, reaction time, heating rate, absence or presence of oxygen, use of catalysts, and pressure, affect the distribution and quality of the products generated (Capareda 2014; Maguyon and Capareda 2013; Maguyon-Detras and Capareda 2017). Table 4.1 shows the typical process conditions and products for thermal conversion processes.

Pyrolysis is the thermal conversion of biomass at temperatures typically below 600 °C in the complete absence of oxygen to produce higher-energy density materials including char, bio-oil, and gaseous products. Gasification usually occurs at temperatures greater than 600 °C with an oxygen supply that is only a fraction of what is theoretically required for complete combustion. The resulting gaseous product, which typically consists of hydrogen, carbon monoxide, methane, carbon dioxide and nitrogen, is said to be more versatile than the original feedstock and can be burned to produce process heat and steam or be used in gas turbines to produce electricity (Demirbas 2001). Combustion is the oldest method of thermal conversion of biomass, which accounts for almost 97% of the world's bioenergy production (Demirbas et al. 2009). In this process, the feedstock is subjected to high temperatures (typically above 700 °C) with an excess amount of air to produce gaseous products consisting mainly of CO_2, H_2O, N_2, and heat.

Pyrolysis serves as the core reaction of thermal conversion processes since the irreversible degradation of the biomass starts at temperatures of about 150–200 °C with the absence of oxygen. According to Pollex et al. (2011), even if the feedstock is surrounded by oxygen from the air, it does not participate in the reaction since the pyrolysis gases produced during this process flow out from the biomass particles and prevent oxygen from reaching it. As the temperature goes up and some oxygen (less

Table 4.1 Overview of thermal conversion processes, their process conditions and products

	Pyrolysis	Gasification	Combustion
Process conditions			
Temperature (°C)	300–600	>600	>700
Reaction time	1 s (fast pyrolysis) days (slow pyrolysis)	Several seconds to minutes	–
Air supply, λ[a]	$\lambda = 0$	$\lambda = 0.2–0.5$	$\lambda \geq 1$
Products			
Gaseous product	CO, CH_4, C_XH_Y, CO_2, H_2O, pyrolysis oils, N- and S-containing compounds	CO, H_2, CO_2, H_2O, CH_4, C_XH_Y, tars, NH_y, NO_x, H_2S, COS	CO_2, H_2O, CO, C_XH_Y, NO_X, SO_X
Solid	C_m H_n O_k, (N, S), ash	C, (N, S), ash	Ash, (N,S)

Source: Adapted from Capareda (2014) and Lohri et al. (2015)

[a]λ = O_2 supplied/O_2 theoretical

than the theoretical oxygen) is introduced, the process shifts to gasification. Any further increases in temperature and oxygen supply lead to combustion reactions.

The composition of the products produced during thermal conversion depends on the severity of process conditions and oxygen supply. Thermal processes tend to shift the biomass components into different materials in terms of their carbon, hydrogen, and oxygen contents. Slow pyrolysis, for example, tends to increase the carbon content of the feedstock producing char, while rapid or fast pyrolysis produces more hydrocarbons. Oxidation, on the other hand, produces gaseous products, such as CO_2 and H_2O. Steam and hydrogasification tend to produce gaseous products enriched with hydrogen.

4.2 Properties of Rice Straw for Thermal Conversion

Biomass properties and chemical composition are important in determining a material's suitability for thermal conversion. Chapter 1 presented rice straw properties and composition. The *moisture content* of the biomass is one of the characteristics initially considered and is an important criterion in selecting the appropriate technology for conversion. Biomass feedstocks with moisture contents usually above 60 to 65%, by weight, have very low calorific values to be considered for combustion. Hence, biogas production is more feasible if drying is not considered (IFC 2017). For gasification, moisture contents of about 50 and 60%, by weight, can be handled using updraft bed gasifiers and fluidized bed gasifiers, respectively (Roos 2009). For pyrolysis, the moisture content of the biomass should be below 20–25% (Dong et al. 2016). Some reactor designs, however, may be amenable to higher moisture contents such as the multi-sectional rotary kilns used for pyrolysis of wastes with high organic content (Chen et al. 2015). As mentioned in Chap. 1, however, rice straw moisture content widely varies depending on the handling, collection and storage methods, and duration.

Physical properties such as particle size, specific heat capacity, and thermal conductivity of the biomass, on the other hand, affect the rate at which heat and oxygen penetrates the biomass particles during thermal processing. *Particle size or particle size distribution (PSD)* is the measure of the physical dimensions of the biomass material and can be obtained using various standard sieves. Biomass materials (i.e., agricultural residues) vary in sizes and need to be ground to less than 10 mm in size for various conversion processes (Capareda 2014). Loose rice straw is typically long and needs to be chopped into smaller pieces. To do so, an additional chopper or shredder is needed, entailing additional energy input for the process. Particle size is particularly important in the pyrolysis process for it can control the rate of heat transfer in the biomass, making it a major factor in the rate of drying and primary pyrolysis reaction (Tripathi et al. 2016; Isahak et al. 2012; Demirbas 2004). *Specific heat capacity* is defined as the ratio of the amount of heat energy transferred to or from the material to the resulting increase in temperature of this material per unit mass (i.e., $J\ kg\text{-}K^{-1}$). *Thermal conductivity*, on the other hand, is the ability of the material to conduct or transfer heat (expressed in W m-K^{-1}) (Capareda 2014). According to Czernik (2010), biomass with large particle size has low thermal conductivity that results in slow heat and a mass transfer rate.

The chemical composition of the biomass may be used to evaluate the thermal degradation of the biomass and residues/pollutants (i.e, ash, NOx, SOx), which may be generated. *Proximate analysis* provides information on the behavior of the feedstock when it is heated (i.e., how much goes off as gas or vapors and how much remains as fixed carbon). It includes the measurement of volatile matter, fixed carbon, and ash. Volatile matter (VM) is the material expelled from the biomass when exposed to 950 °C for 7 min in an oxygen-free environment. It includes volatile carbon, combined water, net hydrogen, nitrogen, and sulfur. VM is important in thermal conversion since high amounts lead to more combustible gases during combustion or more gaseous (condensable and non-condensable) products during gasification and pyrolysis.

Fixed carbon (FC) is the material expelled after burning the moisture- and VM-free biomass to 575 °C for 4 h until the material becomes grayish white (Capareda 2014). FC fraction in the residual biomass usually increases as volatiles are released during thermal degradation (Maguyon and Capareda 2013). Char, the residue after pyrolysis, has typically high amounts of FC and ash. Ash is the residue which remains after complete combustion of the feedstock which can be used for other purposes such as cement aggregate replacement, fertilizer additive, etc. As can be seen in Table 4.2, rice straw FC widely varies among different samples but the range is comparable with other biomass used for thermal conversion such as rice husk and wheat straw. Volatile content is relatively lower compared to other biomass while ash content is relatively higher. Herbaceous biomass such as rice straw has typically higher ash content (up to 25%) compared to woody biomass. This may be attributed to the physiological ash, which results from intrinsic biomass properties such as plant type, maturity and anatomical fractions. High ash contents result in lower energy yields, catalyst impairment, and slag formation during thermal

Table 4.2 Proximate analysis of various biomass

Biomass	Components (% wt)			Sources
	Volatiles	Fixed carbon	Ash	
Rice straw	60.55–78.07	6.93–16.75	14.11–22.70	Migo (2019), Biswas et al. (2017), Fu et al. (2012)
Rice husks	73.41	11.44	15.14	Biswas et al. (2017)
Corn cob	91.16	6.54	2.30	Biswas et al. (2017)
Wheat straw	83.08	10.29	6.63	Biswas et al. (2017)
Coffee hulls	77.50	11.00	11.5	Huang et al. 2011
Bamboo leaves	70.30	18.70	11.00	Huang et al. 2011
Sugarcane bagasse	87.00	4.20	8.80	Huang et al. 2011
Sugarcane peel	77.30	10.10	12.60	Huang et al. 2011

conversion. As such, reducing ash content through washing or leaching may be necessary prior to thermal processing (Kenney et al. 2013).

Ultimate analysis, on the other hand, provides information on the elemental composition of the biomass. It includes carbon, hydrogen, oxygen, nitrogen, sulfur and ash, which can be determined in percent by weight using an elemental analyzer. Results of ultimate analysis can be used to determine the atomic O:C and H:C ratios of the biomass and thermal conversion products (i.e., char), which can be plotted in the van Krevelen diagram to determine their suitability as fuel alternatives (McKendry 2002). Table 4.3 shows that rice straw is within the biomass region due to its high oxygen content. Fossil fuels, such as coal, are near the y-axis in the van Krevelen plot due to high carbon and low oxygen contents. During thermal conversion, the hydrogen and oxygen contents are released from the biomass by volatilization or distillation, tending to increase the carbon content of the residual biomass. The char obtained from pyrolysis, for example, contains higher amounts of carbon as compared to the original biomass. In complete combustion, on the other hand, only the ash remains since all the combustible components (C, H, N, S) of the feedstock are converted to gaseous products (i.e., CO_2, H_2O). Ultimate analysis also shows the potential formation of nitrogen- and sulfur-containing compounds in the products (i.e., NOx and SOx in the gas emissions; nitrogenous compounds in bio-oil) (Maguyon and Capareda 2013, Maguyon-Detras and Capareda 2017). High nitrogen and sulfur contents of the biomass may lead to formation of SOx and NOx during combustion, which causes environmental problems (i.e., acid rain). Pyrolytic oil containing high amounts of nitrogenous and sulfur-containing compounds may need upgrading to meet transport fuel standards. As shown in Table 4.3, rice straw contains minimal amounts of nitrogen and sulfur, which are comparable with other biomass.

Another basis for the suitability of biomass for thermal conversion is its heating value. *Heating value (HV)* or *calorific value (CV)* measures the energy content of the biomass and can be obtained using a calorimeter. As shown in Table 4.4, the higher heating value (HHV) of rice straw is comparable to other biomass materials, such as rice husks and wheat straw indicating its potential as feedstock for thermal conversion.

Table 4.3 Ultimate analysis of various biomass

Component (% wt)	C	H	O	N	S	Sources
Rice straw	45.7–61.4	5.1–8.5	48.3–58.1	0.8–1.4	0.3–0.4	Migo (2019), Fu et al. (2012), Biswas et al. (2017)
Corn cob	54.2	8.2	44.3	2.4	0.6	Biswas et al. (2017)
Wheat straw	54.5	7.6	63	0.6	0.6	Biswas et al. (2017)
Woody	49.1	6.1	44.3	0.4	0	Huang et al. (2011)
Herbaceous	47.8	6.1	45.3	0.8	0.1	Huang et al. (2011)
Ag. residue	46.8	6	45.7	0.8	0.2	Huang et al. (2011)

Table 4.4 Energy content of various biomass

Biomass	HHV (MJ kg^{-1})	Sources
Rice straw	14.2–14.9	Nam et al. (2015), Park et al. (2014), Biswas et al. (2017)
Rice husks	12.9	Biswas et al. (2017)
Corn cob	16.0	Biswas et al. (2017)
Wheat straw	14.7	Biswas et al. (2017)
Sawdust	18.4	Liu et al. (2013)
Cassava stalk	17.6	Pattiya (2011)
Pine wood chips	20.2	Srinivasan et al. (2012)
Sewage sludge	11.4	Abrego et al. (2013)

4.3 Currently Developed Technologies and Practices of Rice Straw Thermal Conversion

Figure 4.1 shows the various routes for bioenergy production in the form of heat, steam, biofuels, or power from rice straw via thermal conversion methods and the potential high-value nonenergy byproducts.

4.3.1 Pyrolysis

Pyrolysis is an irreversible thermal conversion process done at temperatures typically above 300 °C in the complete absence of an oxidant. This process, also known as destructive distillation, degrades biomass components into three primary products: char, bio-oil, and synthesis or producer gas (Bridgwater 2006; Capareda 2014). Several reaction parameters (i.e., temperature, heating rate, residence time, pressure, and catalyst) and biomass type and characteristics (i.e., particle size) greatly affect the proportion and quality of the pyrolysis products (Mahinpey et al. 2009). Table 4.5 shows the process parameters and product yields for various pyrolysis modes including gasification. Generally, more char is obtained at lower temperatures while higher amounts of syngas are produced at higher temperatures due to a higher degree of devolatilization and cracking. The liquid product or bio-oil produced in higher amounts at 500–550 °C may contain various chemicals ranging from aliphatic compounds, acids, alcohols, esters, and nitrogenous compounds (i.e., amines, nitriles) (Chen et al. 2015; Maguyon and Capareda 2013; Maguyon-Detras and Capareda 2017). As shown in Fig. 4.1, all pyrolysis products can be used as fuel.

Char produced from pyrolysis has properties very similar to natural coal and it contains relatively higher energy than the raw biomass (Tag et al. 2016). Bio-oil, on the other hand, has a typically complex composition, high moisture content, and high acidity, hence, physical and chemical upgrading (i.e., column chromatography, distillation, solvent extraction, hydrogenation, and deoxygenation) must be done prior to its application as liquid fuels (i.e., diesel, gasoline) (Teella et al. 2011; Wang

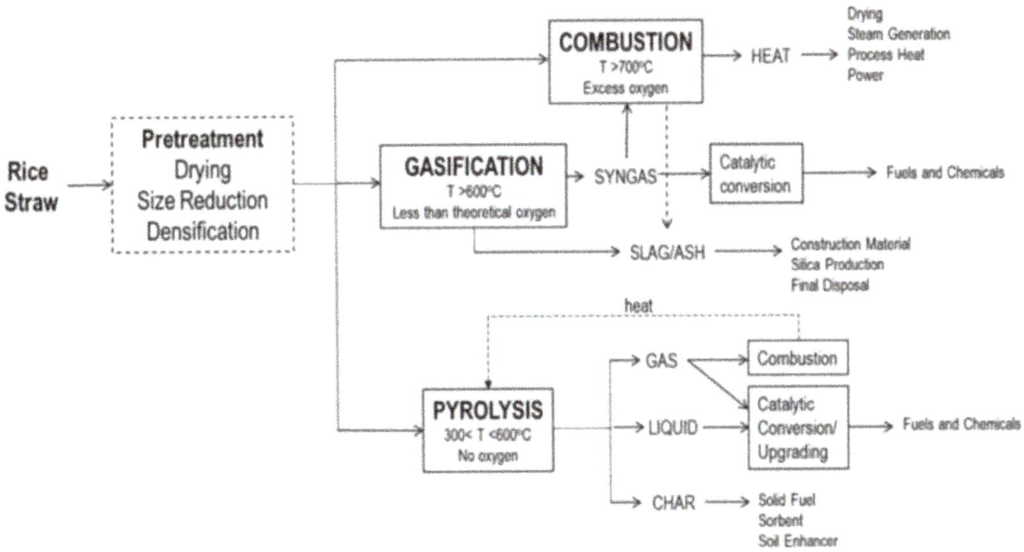

Fig. 4.1 Routes for thermal conversion of rice straw

Table 4.5 Thermal conversion processes, process conditions, and product distribution

Mode	Process conditions		Product distribution (%)		
	Peak temperature	Vapor residence time	Char	Liquid	Gas
Slow	Moderate (~500 °C)	Long (5–30 min)	35%	30% (70% water)	35%
Intermediate	Moderate (~500 °C)	Moderate (10–20 sec)	20–25%	50% (50% water)	25–30%
Fast	Moderate (~500 °C)	Short (< 2 sec)	12%	75% (25% water)	13%
Gasification	High (>800 °C)	Moderate (10–20 sec)	10%	5% tar (55% water)	85%

Sources: Adapted from Bridgwater (2012) and Duku et al. (2011)

et al. 2009, 2011; Cao et al. 2010; Zeng et al. 2011; Amen-Chen et al. 1997; Lu et al. 2011) or for other energy conversions. Syngas, which consists of combustible gases (i.e., methane, hydrogen, carbon monoxide) along with carbon dioxide and nitrogen, is typically recirculated back to the pyrolysis process to supply its heat requirement making the process self-sustaining (Agarwal 2014). According to Chen et al. (2015), energy is produced in a cleaner way via pyrolysis compared to combustion and gasification due to the inert environment producing less NOx and SOx. Also, the syngas produced can be washed before combustion. Pyrolysis systems can also be installed anywhere since their products (bio-oil, char) can be stored and transported making it more flexible than other thermal processes (Agarwal 2014).

Pyrolysis systems generally include facilities for biomass pretreatment (i.e., drying, size reduction), hopper and feeder, pyrolysis reactor, char separation system (i.e., cyclones), and a quenching system for the separation of condensable gases (liquid product) and noncondensable gases (gaseous product). Various reactor configurations can be used for biomass pyrolysis such as rotary kilns, fluidized bed, fixed bed, entrained flow, and tubular reactors. Pyrolysis systems can also be combined with other thermal conversion technologies (Chen et al. 2015).

Data and experience on rice straw pyrolysis are mostly from laboratory-scale experiments. There are limited studies reported on bio-oil production in a bench- or pilot-scale process with capacities ranging from 1 to 3 kg h^{-1} (Park et al. 2004; Tewfik et al. 2011; Yang et al. 2011). Various studies on rice straw pyrolysis show promising results. In the investigation of Nam et al. (2015) using bench scale auger, batch, and fluidized bed reactors, a 43% bio-oil yield from rice straw was obtained using a fluidized bed reactor and 48% bio-char yield was obtained using a batch reactor. The bio-oil and bio-char heating value from the study were 31 and 19 MJ kg^{-1}, respectively, and the energy conversion efficiencies of the different reactors tested ranged from 50 to 64%. In another study (Park et al. 2014) using slow pyrolysis process, the combined energy yields from bio-oil and syngas reached 60% at pyrolysis temperatures over 500 °C. Biochar served as the main product capturing 40% energy and 45% of rice straw carbon. Rice straw was also compared with other lignocellulosic biomass (i.e., corn cob, wheat straw, rice husks) in a study conducted by Biswas et al. (2017). Maximum bio-oil production from rice straw of about 28.4% was observed at 400 °C.

Similar to other lignocellulosic biomass, rice straw, which contains about 38.3% cellulose, 22.2% hemicellulose, and 14.23% by weight lignin (Ukaew et al. 2018) exhibits maximum decomposition at temperatures ranging from 300–450 °C (Biswas et al. 2017). Bio-oil produced using an entrained flow type pyrolyzer at 550 °C contains high carbon content (54.12% wt) and H/C molecular ratio contributing to high energy content of about 29 MJ kg^{-1} (Tewfik et al. 2011). Rice straw bio-oil, however, was found to be slightly acidic (Yang et al. 2011; Park et al. 2004). This can be improved by pretreating the biomass by torrefaction prior to pyrolysis (Ukaew et al. 2018). The removal of minerals via dilute acid washing can also increase bio-oil production, particularly levoglucosan. Baloch et al. (2016) also suggested that leaching rice straw with water can improve the pyrolysis process. Higher bio-oil yields at temperatures greater than 500 °C were also obtained from acid-washed rice straw compared to untreated rice straw using a fixed bed reactor operated at 300 to 700 °C (Li et al. 2012). Some drawbacks on rice straw pyrolysis include the need for biomass drying and grinding, which requires additional energy input to the process.

4.3.2 *Gasification*

Gasification is the thermal conversion of carbonaceous biomass in an oxygen-deficient environment to produce synthesis gas (or syngas) through a series of chemical reactions. The basic reactions involved in gasification are the following (Agarwal 2014; Young 2010):

Partial oxidation:	$C + ½ O_2 \leftrightarrow CO$	$\Delta H = -268$ MJ kmol^{-1}
Complete oxidation:	$C + O_2 \leftrightarrow CO$	$\Delta H = -406$ MJ kmol^{-1}
Water gas reaction:	$C + H_2O \leftrightarrow CO + H_2$	$\Delta H = +118$ MJ kmol^{-1}
Water gas shift reaction:	$CO + H_2O \leftrightarrow CO_2 + H_2$	$\Delta H = -42$ MJ kmol^{-1}
Steam methane reforming:	$CH_4 + H_2O \leftrightarrow CO + 3H_2$	$\Delta H = +88$ MJ kmol^{-1}
Hydrocarbon reactions:	$C_nH_m + nH_2O \leftrightarrow nCO + (n + m/2)H_2$	(Endothermic)

Syngas is mainly composed of combustible gases, such as CO and H_2, and its composition varies with process conditions (i.e., temperature, pressure), reactor design, feedstock characteristics, and gasifying agent (air, steam, oxygen) (Agarwal 2014). Syngas treatment processes aim to further increase combustible components (i.e., H_2, CO, C_xH_y) by removing noncombustible gases and water. Syngas can be used as an energy source for heating, drying, cooking, biofuel production, or as a cogeneration system to produce electricity. It can also be used as a feedstock for the manufacture of high-value chemical compounds. According to Young (2010), the CO and H_2 in syngas serve as building blocks for the synthesis of various industrial chemical compounds including methanol, hydrogen, and ammonia among others. Fuels in the form of alcohols and diesel can also be produced from syngas via the Fisher-Tropsch method. The inorganic materials present in the biomass are converted into a solid rock-like material referred to as slag or vitrified slag or ash during gasification.

The technologies for gasification mostly evolved from the gasification of coal since it was one of the first technologies for syngas production. Generally, there are three basic types of reactors used for gasification: (1) moving-bed or fixed-bed gasifier, (2) fluidized-bed gasifier, and (3) entrained-flow gasifier. Table 4.6 summarizes the variation among these three gasification configurations.

Studies on rice straw gasification show initial positive results for bioenergy production. Gasification studies using rice straw in fluidized bed gasifier (to produce syngas) resulted in 61% hot gas efficiency and 52% cold gas efficiency, with the higher heating value of about 5.1 MJ N m^{-3} (Calvo et al. 2012). Bed agglomeration was solved by substituting the usual alumina-silicate bed by a mixture of alumina-silicate sand and magnesium oxide (MgO) (Calvo et al. 2012). Another study on gasification showed that the presence of potassium carbonate (K_2CO_3) improved the production of H_2-rich gas with yields up to 59.8% H_2 (Baloch et al. 2016).

The single largest problem in gasification is the occurrence of tar in the producer gas, which requires strategies for dealing with the tar either by removal using filters, scrubbers, or condensers, or by in situ conversion through catalytic cracking or reforming of tar, both of which are still under development (Brandin et al. 2011).

Table 4.6 Process description and application of various gasification types

Type	Process description	Other characteristics	Outlet gas temp. (°C)	Technology providers	Application
Fixed or moving bed	Feedstock enters at the top while steam/air enters at the bottom; syngas exits from the top; slag or ash is removed from the bottom; may be operated below or above ash-slagging temperature range	Low oxygen requirement; tars and oils are produced; limited ability to handle fine feed; slagging and nonslagging ash depending on operating temperature	430–650	British gas/Lurgi (BGL) and Lurgi	42% of total installed gasification capacity worldwide
Flui-dized bed	Feedstock enters the side of the reactor while steam/air enters at the bottom; syngas exits from the top; ash is removed from the bottom; steam/air fluidize the feed particles; temperature is maintained below the ash fusion range	Moderate oxygen/ steam requirement; nonslagging ash; can process a wide range of feedstock; bed temperature is uniform and moderate; extensive char recycling	930–1040	HTW, KRW, KBR and Winkler	2% of total installed gasification capacity worldwide
En-trained flow	Feedstock and steam/air enters at the top; syngas exits near the bottom; ash is removed from the bottom; operates at temperatures above ash-slagging conditions	Reliable and proven design; no internal moving parts; compact size; minimal byproducts; has ability to supply syngas at high pressures; uniform temperature; short residence time; slagging ash; feed should be finely divided and homogenous; large oxygen requirement; high temperature slagging operation; possible entrainment of molten slag in the raw syngas	1230–1650	ConocoPhillips, future energy, GE Energy, and Shell	56% of total installed gasification capacity worldwide

Source: Adapted from Young (2010)

4.3.3 Combustion

Combustion is the confined and controlled process of burning organic materials (i.e., biomass, wastes) with sufficient amount of air (usually 110 to 150% of stoichiometric oxygen) at temperatures typically between 700 and 1350 °C to produce heat, mechanical, or electrical energy using various equipment such as stoves, furnaces, boilers, etc. (Tan 2013). Any kind of waste material with moisture content of <50% can be used as fuel for combustion (Agarwal 2014). This process can reduce the volume of waste up to about 90% (70% by mass) depending on the feedstock and reactor types, and process conditions. However, efficient air pollution control devices are required for treatment of the flue gases, which may contain various air pollutants (i.e. NOx, SOx, PMs, dioxins). Bottom ash and fly ash are also produced from the inorganic fraction of the fuel used. The scale of a combustion plant may range from domestic-level heating to large industrial plant. The complete combustion reaction of hydrocarbons is shown below:

$$\mathrm{C_nH_m} + (n + m/4)\mathrm{O_2} \rightarrow nCO_2 + (m/2)\mathrm{H_2O}$$

Unlike for rice husks, research and applications of bioenergy via thermal conversion of rice straw are in developmental stage. And among the different thermochemical processes, direct combustion is the most established because of its simplicity and high efficiency; thus, it can be used for the economic heat generation from biomass (Nussbaumer 2003). A study conducted at IRRI in 2018 used a bench scale, direct combustion rice straw furnace (Fig. 4.2a) to heat air for paddy drying using a flatbed dryer. Results from the study show that at feed rates of 20 to 30 kg h^{-1} rice straw, the energy output of the furnace ranged from 200 to 350 MJ h^{-1}; and the drying air temperature is raised from 14 to more than 30 °C above ambient, proving to be sufficient for paddy drying applications (Migo 2019). In addition, drying air efficiency ranged from 60 to over 85% in the bench scale application (Migo 2019). However, due to the ash accumulation problem, manual feeding was necessary to break the ash (Fig. 4.2b).

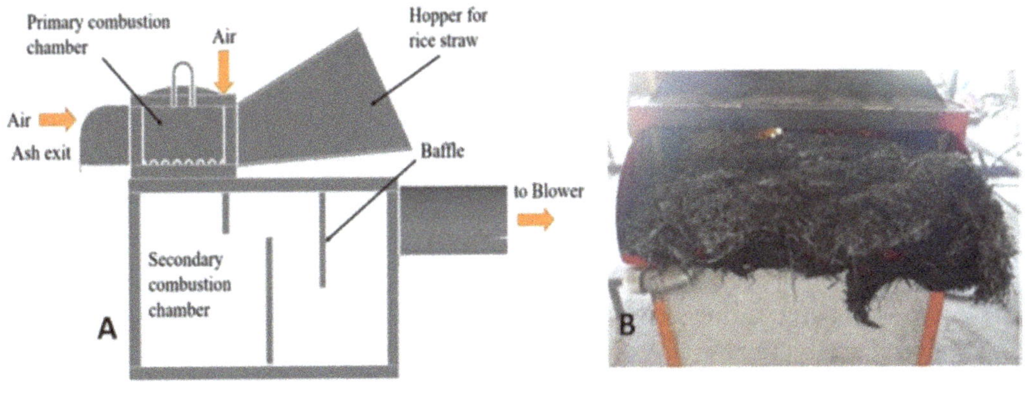

Fig. 4.2 (**a**) Diagram of rice straw furnace and (**b**) photo of refuse ash

There are many ash-related issues encountered in combusting rice straw and other herbaceous biomass including: accumulation, slagging, fouling, and corrosion of the boiler due to the chlorine and alkali content (Zafar 2018). Although there are no reported large-scale direct combustion power plants using rice straw as of 2018, large-scale application could be highly feasible—as demonstrated by combined heat and power plants (CHPs) in Europe that operate using wheat and oat straw. Traditionally, European power plants feed straw bales directly in combustion chambers (also known as "Vølund cigar feeding"), to save on additional energy input for dissolving the bales; however, more advanced power plants use shredded straw, combining it with coal (i.e., coal and straw co-firing) in fluidized bed systems (Zafar 2018) for higher efficiencies. Thus, rice straw pretreatment may be necessary before energy conversion but may have a negative effect in the overall energy balance.

In terms of pretreatment, particle size may be reduced after 2 weeks of air drying. Chou et al. (2009) reported that rice straw should be reduced to about 2–5 mm length for energy conversion. Rice straw may also be compressed into cubes with sides of 50 mm^3 using hot press machines to increase its density and energy content for a more uniform combustion (Kargbo et al. 2010). In order to solve slagging and fouling problems, leaching may be used as a pretreatment method. In a study by Kargbo et al. (2010), pretreatment was done by spraying water on spread rice straw (not more than 30 cm thick) over a steel mesh, which resulted in reduced concentration of K, Na, and Cl.

The size of currently available biomass combustion systems ranges from a few kW up to more than 100 MW (Nussbaumer 2003). Biomass combustion systems can be classified according to capacity: small–scale (less than 100 kWth), medium-scale (from 100 kWth to 10 MWth), and large-scale (greater than 10 MWth) (Obernberger 2008). However, because of logistical problems in gathering rice straw and transporting it to power plants, small-scale rice straw bioenergy applications near farming fields are more common.

4.4 Nonenergy Thermal Conversion Byproducts

This section focuses on the nonenergy applications of thermal conversion products, such as biochar from pyrolysis and ash residues from gasification and combustion systems.

4.4.1 Biochar

Biochar is the charcoal-like carbon-rich residue produced during biomass pyrolysis. Generally, it must contain at least 50% wt carbon, 75% of which is fixed carbon (Jonsson 2016). Due to the release of volatile matter from the biomass during pyrolysis, biochar contains relatively high carbon and lower oxygen content than the raw

Fig. 4.3 (**a**) Rice straw and (**b**) rice straw-derived biochar

biomass, which is desirable especially if a high heating value is needed. The low amount of sulfur and nitrogen in biochar is also favorable in preventing high emissions of sulfur and nitrogen oxides into the atmosphere (Liu and Han 2015) (Fig. 4.3).

Biochar is used in agriculture to condition the soil, increase crop production, and mitigate GHG emissions. It can serve as a water reservoir due to its high water-retention capacity (Schimdt 2012). Also, it can bind nutrients into its structure, thus, slowing the rate of nutrient loss (Amarasinghe et al. 2016). Biochar has generally a neutral to alkaline pH, hence, it can also be applied to adjust the pH value of acidic soils (Zhang et al. 2015; Ahmad et al. 2014). In a study conducted by Wu et al. (2012), rice straw was used as a feedstock for slow pyrolysis to determine the yields and characteristics of biochar produced at various temperatures and residence times. Based on their study, rice straw biochar has high surface alkalinity, high cation exchange capacity, and contains high amounts of macronutrients (i.e., phosphorus, potassium) indicating its suitability as soil enhancer. The biochar produced at higher pyrolysis temperatures is more carbon-enriched and contains aromatic compounds, which may be recalcitrant in soil improving its capability for carbon sequestration (Thammasom et al. 2016; Wu et al. 2012).

Increased carbon storage in soil with biochar was also reported by Yun-Feng et al. (2014), which may be due to higher aryl- and carbonyl-C contents of rice straw biochar. Yang et al. (2019) studied the effect of applying rice straw biochar in paddy fields under controlled irrigation. Their results indicate that biochar loading at 20 and 40 t ha^{-1} can reduce CH_2 and N_2O emissions, increase rice yields, and improve irrigation water productivity under controlled irrigation. Qin et al. (2016) reported the same results for 20-t-ha^{-1} biochar loading in rice fields, which resulted to maximum GHG-emission reduction of about 36.24% compared to a traditional field management method involving chemical fertilizer application. The reduction in GHG emissions was attributed to the increase in biodiversity and abundance of methanotrophic microbes, increased soil pH, increased soil aeration, and the recalcitrance of biochar.

Biochar can also be used in treating various organic and inorganic contaminants, such as heavy metals, herbicides, and antibiotics in soil and aqueous solutions (Tan et al. 2015. It is comparable to activated carbon, which is the most widely used adsorbent. But unlike activated carbon, biochar produced at certain pyrolysis conditions can be used directly without activation due to partitioning in the noncarbonized fractions or electrostatic attraction with O-containing carboxyl, hydroxyl, and phenolic surface functional groups which could effectively bind with contaminants (Uchimiya et al. 2012).

Generally, the functionality of biochar as a sorbent varies with pyrolysis temperature since the release of O- and H-containing functional groups changes its surface polarity and aromaticity. Different types of contaminants also have varying affinities with biochar. Organic contaminants can be removed from the soil and aqueous solutions through adsorption, partitioning in the noncarbonized fraction, or electrostatic attraction while inorganic pollutants can be removed by physical adsorption, ion exchange, electrostatic attraction, or precipitation (Ahmad et al. 2014). Rice straw-derived biochar has also been found to be effective in treating various soil and aqueous pollutants including pentachlorophenol, dyes, and lead (Lou et al. 2011; Qiu et al. 2009; Jiang et al. 2012).

4.4.2 Slag, Vitrified Slag or Ash

Solid rock-like material, referred to as slag or vitrified slag, and greyish residue called ash are considered to be byproducts of the gasification and combustion processes. These materials are derived from the inorganic fraction of the biomass. Process conditions, such as temperature under which the inorganic materials from the feedstock are being converted, determine the property of this material as to whether a slag, vitrified slag, or ash will be formed. Heavy metals including chromium, lead, and mercury, which are considered as environmental hazards, may be present in the slag, vitrified slag, or ash. Hence, the leachability of these metals from the material should be tested. Slag, vitrified slag, and ash are typically disposed of in landfills but these can potentially be converted into high-value products. Due to its rock-like characteristics, slag or vitrified slag may be used as construction material, aggregates in asphalt or cement-concrete, pipe bedding material, decorative tiles, and others if the Toxicity Characteristic Leaching Procedure (TCLP) standards are being met. Ash, on the other hand, may be used as an additive for cement manufacturing (Young 2010).

Ash from gasification or combustion operations may also be utilized for cement and concrete production, road pavement, glasses and ceramics, fertilizers, stabilizing agent,, and zeolite production (Lam et al. 2010). Furthermore, rice straw ash has a high SiO_2 content ranging from 67.68 to 82.6% by weight as discussed in Chap. 1. High silica content of about 73.65% SiO_2 of rice straw ash was also reported by Chen et al. (2017). Silica from ash can be used to treat effluents containing heavy metals (i.e. Pb^{2+}, Cu^{2+}, Cd^{2+}, and Cr^{2+}). Silica extraction from combustion ash has already been studied by Caraos (2018) and Liu et al. (2014) via alkaline fusion and hydrothermal desilication methods.

4.5 Comparison of Thermal Conversion Technologies

Comparisons of thermal conversion methods will be discussed in this section based on energy conversion, technological maturity, and potential environmental issues that may arise from their operation.

4.5.1 *Energy Conversion*

The energy conversion efficiency, or thermal efficiency, is the ratio of the energy generated from the fuel using the selected thermal conversion process and the net calorific value of the fuel. The energy conversion efficiencies of thermal processes are summarized in Table 4.7. On average, about a 60% energy conversion efficiency can be achieved using rice straw as a biomass feedstock. In the study conducted by Nam et al. (2015), energy recovery was accounted for the energy contained in the three pyrolysis products. Higher energy recovery was recorded for char in the bench-scale auger-type and fixed-bed pyrolyzer, while in the fluidized-bed reactor bio-oil was the main energy product. Park et al. (2014) also reported that char contained most of the biomass energy recovered under slow pyrolysis. The energy efficiency for gasification, on the other hand, ranged from 52 to 61%, which is still comparable with other thermal technologies. Lower energy yield was reported by Darmawan et al. (2017); however, this result was obtained using a simulated process only.

Table 4.7 A comparison of thermochemical conversion technologies using rice straw

Thermal conversion process	Scale	Rice straw feed rate	Energy conversion efficiency, %	Reactor temperature, °C	Sources
Direct combustion	Bench scale	20–30 kg h^{-1}	60–85	300–500	Migo (2019)
Pyrolysis (using batch and fluidized bed reactors)	Bench scale	N/A	64.6–75	500	Nam et al. (2015)
Slow pyrolysis	Lab scale	N/A	80–90	>400	Park et al. (2014)
Gasification (using fluidized bad reactor)	Bench scale	3.6–5.4	52–61	700–850	Calvo et al. (2012)
Gasification (using entrained flow and torrefied biomass)	Simulated process	12.4 t/h	43	1200	Darmawan et al. (2017)
Gasification (in the presence of K_2CO_3)	Lab scale	N/A		750	Baloch et al. (2016)

4.5.2 Technology and Commercial Maturity

Direct combustion technology is by far the most technologically developed and commercially available thermal process for biomass. According to Tan (2013), commercial availability represents the market availability of the technology at a commercial scale capacity (at least 25 MW or 300 t day^{-1}). Although there have been a few studies written on large-scale combustion of rice straw for power generation, Boucher et al. (2013) reported that there are at least three power plants producing 24–25 MW in China, which are known to use rice straw as part of the biomass fuel and around 15 biomass power projects in operation by the end of 2012. Some power plants using rice straw as feedstock were also installed in India and Thailand; however, problems on straw collection and feeding especially at large-scale operations led to their closure. This experience suggests that small combustion systems similar to the farm-scale paddy flatbed dryer developed at IRRI, which can be placed near rice fields, may be more feasible (Boucher et al. 2013; Migo 2019). Direct combustion is also highly favored by industrialized countries compared to other conversion technologies due to its lower annual capital cost, operational cost, and higher daily throughput.

Gasification, on the other hand, is relatively more advanced in terms of technological status as compared to pyrolysis. While no commercial rice straw gasification plants have been installed, other biomass materials, such as wood chips or rice husks, have been used for more than 100 years. Fluidized bed systems are in its demonstration phase while downdraft and updraft gasifiers are already in their early commercial stages (IFC 2017). Ouda et al. (2016) also reported that Asia has advanced greatly in the gasification technology over the past few years and can be considered as one of the most favorable markets for gasification technology followed by Europe, Africa, and USA.

Brandin et al. (2011), for example, assessed five gasification plants using different biomass materials as fuel in Europe with electrical power outputs in the range of from 17.5 kWe to 6 MWe and concluded that the technology is generally sufficiently mature for commercialization, although some unit operations, for example catalytic tar reforming, still need further development. All plants have a fully automatic operation and are equipped with complex tar removal systems. Since these are all pilot plants, data about commercial feasibility were not yet available. Further development is also needed to increase the biofuel-to-electricity efficiency, currently at from 20–25%, to 30–40% and overall performance efficiency to around 90%.

Dimpl (2010) analyzed experiences with small-scale applications of the gasification technology worldwide over the last three decades. The study concluded that there is not yet any standard gasifier technology complying with environmental standards appropriate for rural small-scale applications readily available off the shelf. In general, the small-scale power-gasifier technology proved to be unreliable and expensive and to minimize cost systems for capturing carcinogenic waste are often not installed or not very effective.

4.5.3 Environmental and Health Impacts

Among the thermal conversion technologies, pyrolysis results in the least emissions since all products can be used as energy sources or marketable nonenergy products. Studies also show that carbon can be sequestered in the soil by using char as a soil enhancer or fertilizer thus reducing GHG emissions (Roos 2009; Ahmad et al. 2014). Moreover, the use of char as fertilizer can reduce the need for chemical fertilizers and increase nutrient uptake efficiency. However, these positive impacts may be offset if the energy input to the process will come from fossil fuels. This provides motivation for the recirculation of heat from syngas combustion as suggested by Agarwal (2014). Upgrading of the bio-oil, on the other hand, may render it as a comparable replacement to petroleum-derived liquid transport fuels.

Compared to combustion, gasification has lower carbon and NOx emissions since NOx are produced at temperatures higher than the gasification range. Combustion only produces heat, while gasification produces excess heat and gaseous fuel, which can be used in a reciprocating engine, gas turbine, fuel cells or in an integrated gasification combined cycle resulting to higher energy efficiency and lower GHG emissions (Roos 2009). Also, direct combustion may result in emission of toxic substances such as dioxins and furans especially if chlorinated compounds are present in the feedstock. Untreated rice straw has about 0.4% Cl (Roos 2009). Pyrolysis, on the other hand, has the advantage of producing gaseous product that is free from such pollutants due to its inert atmosphere (Agarwal 2014).

More attention is needed by the research community to find solutions to potential health hazards from small-scale gasification. To minimize investment costs to make gasification economically viable, often simple filters are installed for tar removal in small gasification plants, which often produce much carcinogenic waste, especially in the case of wet stripping of the gas. This causes severe environmental and health threats (Dimpl 2010). None of the small-scale plants that were repeatedly monitored in studies since 1995 (Stassen 1995; Dimpl 2010, IRRI unpubl. Trip reports from Myanmar and Cambodia, 2008–13) took adequate measures to deal with the condensates; instead the pollutants were freely discharged into the environment. In addition, the operators dealing with these contaminated condensates often did not use protective clothing or gloves and some complained about frequent headaches.

Shackley et al. (2011) studied soil improvement and carbon sequestration using gasification biochar from rice husk gasifiers installed in Cambodian rice mills and ice factories and highlight that questions remain regarding the safety of the biochar for human health.

4.6 Conclusions and Recommendations

This chapter shows the potential of rice straw in producing bioenergy in the form of biofuels, heat, steam, or power using thermal conversion technologies. Pyrolysis, gasification, and combustion are at different stages of development. Combustion systems for biomass are commercially available while gasification and pyrolysis are still in the demonstration and research stages, respectively. These processes show potential for large-scale application; however, problems associated with rice straw collection, storage, and transportation makes small- or field-scale application a more viable alternative. Further research and development is needed on gasification in the areas of gas cleaning/upgrading, utilization of produced heat to increase overall efficiency and system integration/optimization.

Compared to pyrolysis or gasification, combustion systems are more technologically mature but proper air pollution control devices should be part of these installations to curtail release of harmful gases. Pyrolysis, on the other hand, has lesser emissions due to its inert environment and the residual product—biochar—can be used for carbon sequestration. However, more studies are needed for the feasibility of pyrolysis of rice straw on a commercial scale. In general, the thermal conversion technologies can recover energy from rice straw up to about 60% and the marketable non-energy by-products can potentially improve process economics.

References

Abrego J, Sanchez JL, Arauzo J, Fonts I, Gil-Lalaguna N, Atienza-Martinez M (2013) Technical and energetic assessment of a three-stage thermochemical treatment for sewage sludge. Energy Fuel 27:1026–1034

Agarwal M (2014) An investigation on the pyrolysis of municipal solid waste. Unpublished thesis. RMIT

Ahmad M, Rajapaksha AU, Lim JE, Zhang M, Bolan N, Mohan D, Vithanage M, Lee SS, Ok YS (2014) Biochar as a sorbent for contaminant management in soil and water: a review. Chemosphere 99:19–33

Amarasinghe HAHI, Gunathilake SK, Karunarathna AK (2016) Ascertaining of optimum pyrolysis conditions in producing refuse tea biochar as a soil amendment. Procedia Food Sci 6:97–102

Amen-Chen C, Pakdel H, Roy C (1997) Separation of phenols from eucalyptus wood tar. Biomass Bioenergy 13:25–37

Baloch HA, Yang T, Sun H, Li J, Nizamuddin S, Li R et al (2016) Parametric study of pyrolysis and steam gasification of rice straw in presence of K_2CO_3. Korean J Chem Eng 33(9):2567–2574. https://doi.org/10.1007/s11814-016-0121-7

Biswas B, Pandey N, Bisht Y, Singh R, Kumar J, Bhaskar T (2017) Pyrolysis of agricultural biomass residues: comparative study of corn cob, wheat straw, rice straw, and rice husk. Bioresour Technol 237:57–63

Boucher E, Guillemot A, Pasquiou V (2013) Feasibility study for the implementation of two ORC power plants of 1 MWe each using rice straw as fuel in the context of a public-private partnership with the institutions PhilRice and UPLB. Enertime, Puteaux

Brandin J, Tunér M, Odenbrand I (2011) Small-scale gasification: gas engine CHP for biofuels. Swedish Energy Agency report, Växjö/Lund. Linnaeus University and Lund University

Bridgwater T (2006) Review: biomass for energy. J Sci Food Agr 86:1755–1768

Bridgwater A (2012) Review of fast pyrolysis of biomass and product upgrading. Biomass Bioenergy 38:68–94. https://doi.org/10.1016/j.biombioe.2011.01.048

Calvo L, Gil M, Otero M, Morán A, García A (2012) Gasification of rice straw in a fluidized-bed gasifier for syngas application in close-coupled boiler-gasifier systems. Bioresour Technol 109:206–214. https://doi.org/10.1016/j.biortech.2012.01.027

Cao J-P, Zhao X-Y, Morishita K, Wei X-Y, Takarada T (2010) Fractionation and identification of organic nitrogen species from bio-oil produced by fast pyrolysis of sewage sludge. Bioresour Technol 101:7648–7652

Capareda SC (2014) Introduction to biomass energy conversions. CRC, Boca Raton

Caraos NAP (2018) Parametric analysis of silica extraction process via alkaline fusion and hydrothermal desilication methods from municipal solid waste bottom ash derived from direct combustion. Unpubl. University of the Philippines Los Baños

Chen D, Yin L, Wang H, He P (2015) Pyrolysis technologies for municipal solid waste: a review. Waste Manag 37:116–136

Chen T, Cao J, Jin B (2017) The effect of rice straw gasification temperature on the release and occurrence modes of Na and K in a fluidized bed. Appl Sci 7(12):1207

Chou C, Lin S, Lu W (2009) Preparation and characterization of solid biomass fuel made from rice straw and rice bran. Fuel Process Technol 90(7–8):980–987. https://doi.org/10.1016/j.fuproc.2009.04.012

Czernik S (2010) Fundamentals of charcoal production. National Renewable Energy Laboratory (NREL), Newcastle

Darmawan A, Fitrianto AC, Aziz M, Tokimatsu K (2017) Enhanced electricity production from rice straw. Energy Procedia 142:271–277

Demirbas MF (2001) Biomass resource facilities and biomass conversion processing for fuels and chemicals. Energy Convers Manag 42:1357–1378

Demirbas A (2004) Effect of temperature and particle size on biochar yield from pyrolysis of agricultural residues. J Anal Appl Pyrolysis 721:243–248

Demirbas MF, Balat M, Balat H (2009) Potential contribution of biomass to the sustainable energy development. Energy Convers Manag 50:1746–1760

Dimpl E (2010) Small-scale electricity generation from biomass. Part I: biomass gasification, 1st edn. GTZ-HERA. Poverty-oriented Basic Energy Service

Dong J, Chi Y, Tang Y, Ni M, Nzihou A, Weiss-Hortala E, Huang Q (2016) Effect of operating parameters and moisture content on municipal solid waste pyrolysis and gasification. Energy Fuel 30(5):3994–4001

Duku MH, Gu S, Hagan EB (2011) Biochar production potential in Ghana: a review. Renew Sustain Energy Rev 15:3539–3551

Fu P, Hu S, Xiang J, Sun L, Su S, Wang J (2012) Evaluation of the porous structure development of chars from pyrolysis of rice straw: effects of pyrolysis temperature and heating rate. J Anal Appl Pyrolysis 98:177–183

Huang YF, Kuan WH, Chiueh PT, Lo SL (2011) Pyrolysis of biomass by thermal analysis-mass spectrometry (TA-MS). Bioresour Technol 102:3527–3534

IFC (International Finance Corporation) (2017) Converting biomass to energy: a guide for developers and investors. Retrieved 10 July 2018 https://www.ifc.org/wps/wcm/connect/7a1813bc-b6e8-4139-a7fc-cee8c5c61f64/BioMass_report_06+2017.pdf?MOD=AJPERES

Isahak NRW, Hisham MWM, Yarmo MA, Hin TY (2012) A review on bio-oil production from biomass by using pyrolysis method. Renew Sust Energ Rev 16:5910–5923

Jiang T-Y, Jiang J, Xu R-K, Li Z (2012) Adsorption of Pb(II) on variable charge soils amended with rice-straw derived biochar. Chemosphere 89(3):249–256

Jonsson E (2016) Slow pyrolysis in Brista. Retrieved March 30, 2017 from www.diva-portal.org

Kargbo F, Xing J, Zhang Y (2010) Property analysis and pretreatment of rice straw for energy use in grain drying: a review. Agric Biol J N Am 1(3):195–200. https://doi.org/10.5251/abjna.2010.1.3.195.200

Kenney KL, Smith WA, Greasham GJ, Westover TL (2013) Understanding biomass feedstock variability. Biofuels 4(1):111–127

Lam CHK, Ip AWM, Barford JP, McKay G (2010) Use of incineration MSW ash: review. Sustainability 2:1943–1968. https://doi.org/10.3390/su2071943

Li S, Yu S, Wang F-C, Wang J (2012) Pyrolytic characteristics of rice straw and its constituents catalyzed by internal alkali and alkali earth metals. Fuel 96:586–594

Liu Z, Han G (2015) Production of solid fuel biochar from waste biomass by low temperature pyrolysis. Fuel 158:159–165

Liu W-J, Tian K, Jiang H, Zhang X-S, Yang G-X (2013) Preparation of liquid chemical feedstocks by co-pyrolysis of electronic waste and biomass without formation of polybrominated dibenzo-p-dioxins. Bioresour Technol 128:1–7

Liu Z-S, Li WK, Huang C-Y (2014) Synthesis of mesoporous silica materials from municipal solid waste incinerator bottom ash. Waste Manag 34:893–900

Lohri CR, Sweeney D, Rajabu HM (2015) Carbonizing urban biowaste for low-cost char production in developing countries: a review of knowledge, practices and technologies. Joint Report by Eawag, MIT D-Lab and UDSM. https://www.eawag.ch/fileadmin/Domain1/Abteilungen/sandec/publikationen/SWM/Carbonization_of_Urban_Bio-waste/Carbonization_of_Biowaste_in_DCs_FINALx.pdf

Lou L, Wu B, Wang L, Luo L, Xu X, Hou J, Xun B, Hu B, Chen Y (2011) Sorption and ecotoxicity of pentachlorophenol polluted sediment amended with rice-straw derived biochar. Bioresour Technol 102(5):4036–4041

Lu R, Sheng G-P, Hu Y-Y, Zheng P, Jiang H, Tang Y, Yu H-Q (2011) Fractional characterization of a bio-oil derived from rice husk. Biomass Bioenergy 35:671–678

Maguyon MCC, Capareda SC (2013) Evaluating the effects of temperature on pressurized Ppyrolysis of *Nannochloropsis oculata* based on products yields, and characteristics. Energy Convers Manag J 76:764–773

Maguyon-Detras MC, Capareda SC (2017) Determining the operating condition for maximum bio-oil production from pyrolysis of *Nannochloropsis oculata*. Philippine J Crop Sci 42(2):37–47. https://www.cabdirect.org/cabdirect/abstract/20173297237

Mahinpey N, Murugan P, Mani T, Raina R (2009) Analysis of bio-oil, biogas and biochar from pressurized pyrolysis of wheat straw using a tubular reactor. Energy Fuel 23:2736–2742. https://doi.org/10.1021/ef8010959

McKendry P (2002) Energy production from biomass (part 2): conversion technologies. Bioresour Technol 83:47–54

Migo MVP (2019) Optimization and life cycle assessment of the direct combustion of rice straw using a small scale, stationary grate furnace for heat generation. Unpubl. Masters thesis. University of the Philippines Los Baños

Nam H, Capareda SC, Ashwath N, Kongkasawan J (2015) Experimental investigation of pyrolysis of rice straw using bench-scale auger, batch and fluidized bed reactors. Energy 93:2384–2394

Nussbaumer T (2003) Combustion and co-combustion of biomass: fundamental, technologies, and primary measures for emission reduction. Energy Fuel 17(6):1510–1521. https://doi.org/10.1021/ef030031q

Obernberger I (2008) Industrial combustion of solid biomass fuels: state of the art and relevant future developments. [PowerPoint slides]. Retrieved from: https://www.bios-bioenergy.at/uploads/media/Paper-Obernberger-Industrial-combustion-of-solid-bimass-fuels-2008-10-15.pdf

Ouda O, Raza S, Nizami A, Rehan M, Al-Waked R, Korres N (2016) Waste to energy potential: a case study of Saudi Arabia. Renew Sust Energ Rev 61:328–340. https://doi.org/10.1016/j.rser.2016.04.005

Park Y-K, Jeon J-K, Kim S, Kim JS (2004) Bio-oil from rice straw by pyrolysis using fluidized bed and char removal system. Preprints Papers Am Chem Soc Div Fuel Chem 49(2):800–801

Park J, Lee Y, Ryu C, Park Y-K (2014) Slow pyrolysis of rice straw: analysis of products properties, carbon and energy yields. Bioresour Technol 155:63–70

Pattiya A (2011) Bio-oil production via fast pyrolysis of biomass residues from cassava plants in a fluidized-bed reactor. Bioresour Technol 102:1959–1967

Pollex A, Ortwein A, Kaltschmitt M (2011) Thermo-chemical conversion of solid biofuels. Biomass Convers Biorefin 2(1):21–39. https://doi.org/10.1007/s13399-011-0025-z

Qin X, Li Y, Wang H, Liu C, Li J, Wan Y, Gao Q, Fan F, Liao Y (2016) Long-term effect of biochar application on yield-scaled greenhouse gas emissions in a rice paddy cropping system: a four-year case study in South China. Sci Total Environ 569–570:1390–1401

Qiu Y, Zheng Z, Zhou Z, Sheng GD (2009) Effectiveness and mechanisms of dye adsorption on a straw-based biochar. Bioresour Technol 100(21):5348–5351

Roos CJ (2009) Clean heat and power using biomass gasification for industrial and agricultural projects. Northwest CHP Application Center, Olympia

Schimdt HP (2012) 55 uses of biochar. J Terrior-Wine Biodivers (2008). Retrieved September 27, 2016 from http://www.ithaka-journal.net/55-anwendungen-von-pflanzenkohle?lang=en

Shackley S, Carter S, Knowles T, Middelink E, Haefele S, Haszeldine S (2011) Sustainable gasification–biochar systems? A case-study of rice-husk gasification in Cambodia, Part II: field trial results, carbon abatement, economic assessment and conclusions. Energy Policy 41:618–621 https://www.geos.ed.ac.uk/research/subsurface/diagenesis/Sustainable_gasification_biochar_systems._A_case_study_of_rice_husk_gasification_in_Cambodia_gasification__Part_II_Field_trial_results,_carbon_abatement,_economic_assessment_and_conclusions_Energy_Policy__Shackley_Carter_Haszeldine_2012.pdf

Srinivasan V, Adhikari S, Chattanathan SA, Park S (2012) Catalytic pyrolysis of torrefied biomass for hydrocarbons production. Energy Fuel 12:7347–7353. https://doi.org/10.1021/ef301469t

Stassen HE (1995) Small-scale biomass gasifiers for heat and power. A global review. World Bank technical paper 296. Washington, DC http://documents.worldbank.org/curated/en/583671468769505568/Small-scale-biomass-gasifiers-for-heat-and-power-a-global-review

Tag AT, Duman G, Ucar S, Yanik J (2016) Effects of feedstock type and pyrolysis temperature on potential applications of biochar. J Anal Appl Pyrolysis 10:200–206

Tan Y (2013) Feasibility study on solid waste to energy technological aspects. UC Berkeley College of Engineering, Fung Institute for Engineer-ing Leadership

Tan X, Liu Y, Zeng G, Wang X, Hu X, Gu Y, Yang Z (2015) Application of biochar for the removal of pollutants from aqueous solutions. Chemosphere 125:70–85

Teella A, Huber GW, Ford DM (2011) Separation of acetic acid from the aqueous fraction of fast pyrolysis bio-oils using nanofiltration and reverse osmosis membranes. J Membr Sci 378:495–502

Tewfik SR, Sorour MH, Abulnour AMG, Talaat HA, El Defrawy NM, Farah JY, Abdou IK (2011) Bio-oil from rice straw by pyrolysis: experimental and techno-economic investigations. J Am Sci 7(2):59–67

Thammasom N, Vityakon P, Lawongsa P, Saenjan P (2016) Biochar and rice straw have different effects on soil productivity, greenhouse gas emission and carbon sequestration in Northeast Thailand paddy soil. Agric Nat Res 50:192–198

Tripathi M, Sahu JN, Ganesan P (2016) Effect of process parameters on production of biochar from biomass waste through pyrolysis: a review. Renew Sust Energ Rev 55:467–481

Uchimiya M, Bannon DI, Wartelle LH, Lima IM, Klasson KT (2012) Lead retention by broiler litter biochars in small arms range soil: impact of pyrolysis temperature. J Agric Food Chem 60(20):5035–5044. https://doi.org/10.1021/jf300825n

Ukaew S, Schoenborn J, Klemetsrud B, Shonnard DR (2018) Effects of torrefaction temperature and acid pretreatment on the yield and quality of fast pyrolysis bio-oil from rice straw. J Anal Appl Pyrolysis 129:112–122

Wang S, Gu Y, Liu Q, Yao Y, Guo Z, Luo Z, Cen K (2009) Separation of bio-oil by molecular distillation. Fuel Process Technol 90:738–745

Wang Z, Lin W, Song WL, Du L, Li ZJ, Yao JZ (2011) Component fractionation of wood-tar by column chromatography with the packing material of silica gel. Chin Sci Bull 56(14):1434–1441

Wu W, Yang M, Feng Q, McGrouther K, Wang H, Lu H, Chen Y (2012) Chemical characterization of rice straw-derived biochar for soil amendment. Biomass Bioenergy 47:268–276

Yang SY, Wu CY, Chen KH (2011) The physical characteristics of bio-oil from fast pyrolysis of rice straw. Adv Mater Res 328–330:881–886

Yang S, Xiao Y, Sun X, Ding J, Jiang Z, Xu J (2019) Biochar improved rice yield and mitigated CH_4 and N_2O emissions from paddy field under controlled irrigation in the Taihu Lake region of China. Atmos Environ 200:69–77

Young GC (2010) Municipal solid waste to energy conversion processes: economic, technical, and renewable comparisons. Wiley, Hoboken

Yun-Feng Y, Xin-hua H, Ren G, Hong-Liang M, Yu-Sheng Y (2014) Effects of rice straw and its biochar addition on soil labile carbon and soil organic carbon. J Integr Agric 13(3):491–498

Zafar S (2018). Rice straw as bioenergy resource. https://www.bioenergyconsult.com/rice-straw-as-bioenergy-resource

Zeng F, Liu W, Jiang H, Yu H-Q, Zeng RJ, Guo Q (2011) Separation of phthalate esters from bio-oil derived from rice husk by a basification-acidification process and column chromatography. Bioresour Technol 102:1982–1987

Zhang J, Liu J, Liu R (2015) Effects of pyrolysis temperature and heating time on the biochar obtained from the pyrolysis of straw and lignosulfonate. Bioresour Technol 176:288–291

Chapter 5
Anaerobic Digestion of Rice Straw for Biogas Production

Nguyen Vo Chau Ngan, Francis Mervin S. Chan, Tran Sy Nam, Huynh Van Thao, Monet Concepcion Maguyon-Detras, Dinh Vuong Hung, Do Minh Cuong, and Nguyen Van Hung

Abstract Anaerobic digestion (AD) is a process of degradation of organic matter by microorganisms in an oxygen-free environment, which produces biogas, a vital renewable energy source. Using solely an organic source, such as monosubstrates, it is difficult to optimize the AD process due to nutrient imbalance, lack of appropriate microbial communities, and the effect of operational parameters. This chapter reviews the current studies on biogas production from the anaerobic codigestion process of mixing agricultural byproducts, focusing on rice straw and livestock manure as substrates. Because rice straw is high in cellulose, it needs to be pretreated before feeding into the anaerobic digester. Different rice straw pretreatments are summarized including physical, chemical, and biological methods. Current biogas systems are discussed. The utilization of bioslurry from the anaerobic fermentation process to agricultural cultivation and aquaculture activities is also discussed.

Keywords Anaerobic digestion · Biogas · Co-digestion · Pre-treatment · Rice straw

N. V. C. Ngan (✉) · T. S. Nam · H. Van Thao
Can Tho University, Can Tho, Vietnam
e-mail: nvcngan@ctu.edu.vn; tsnam@ctu.edu.vn; hvthao@ctu.edu.vn

F. M. S. Chan
University of the Philippines Los Baños, Los Baños, Philippines
e-mail: f.chan@irri.org

M. C. Maguyon-Detras
Department of Chemical Engineering, College of Engineering and Agro-Industrial Technology, University of the Philippines Los Baños (UPLB), Los Baños, Laguna, Philippines
e-mail: mmdetras@up.edu.ph

D. V. Hung · D. M. Cuong
Hue University, Hue, Vietnam
e-mail: dominhcuong@huaf.edu.vn

N. V. Hung
International Rice Research Institute (IRRI), Los Baños, Laguna, Philippines
e-mail: hung.nguyen@irri.org

M. Gummert et al. (eds.), *Sustainable Rice Straw Management*,
https://doi.org/10.1007/978-3-030-32373-8_5

5.1 Introduction of Anaerobic Digestion Technology

5.1.1 Products of Anaerobic Digestion

Biogas is one of the major products of the anaerobic digestion (AD) of organic substances and is considered an alternative green energy resource. It is a mixture of gases of which the composition depends on substrates and AD process conditions such as temperature, pH, and retention time. Biogas mixture mainly consists of CH_4, CO_2, O_2, N_2, H_2S, and even other gas compositions. Methane (CH_4) is the most important component of biogas because it has the highest energy density among the biogas components. Therefore, the high CH_4 content of biogas is desired. Examples of biogas composition from different substrates or sources are shown in Table 5.1.

Biogas technology is applicable to small-scale and large-scale uses including electricity generation.

Biogas can also be upgraded into biomethane or renewable natural gas (RNG), which is like natural gas that comes from fossil fuels. Its methane content is 90% or greater. RNG can be substituted for natural gas and can be used as fuel for vehicles that run on natural gas and to supply gas to natural gas grid.

Aside from biogas, the other AD byproduct is digestate. It can be solid or liquid and contains considerable amounts of nitrogen (in ammonium form), macronutrients, and micronutrients that can supplement plant growth (Makádi et al. 2012). Depending on the feedstock of AD, digestate can be utilized as an organic fertilizer or as a compost ingredient. Digestate that uses plant and animal-based feedstocks can be used directly as a fertilizer. One example is the VACB model (Vườn/Garden-Ao/Pond-Chuồng/Pigsty-Biogas) of Vietnam (Thanh 2010). Digestate from the AD process that uses an organic fraction of municipal waste as feedstock will need further processing or will need to be disposed of in a landfill because of possible high levels of heavy metals, pathogens, and other toxic substances.

Table 5.1 The composition of biogas obtained from organic substances

Sources	The composition of biogas (%)					Sources
	CH_4	CO_2	O_2	N_2	H_2S	
Landfills	47–57	24–29	< 1	1–17	< 0.1	Rasi et al. (2007)
	37–62	24–29	< 1	–	–	Allen et al. (1997)
	45–55	30–40	–	51–5	–	Jönsson et al. (2003)
Plant biomass	55–58	37–38	<1	<1–2	–	Rasi et al. (2007)
	48–65	36–41	<1	<17	<0.1	Rasi et al. (2007)
	45–54	–	–	–	–	Tran et al. (2014)
Sludge	55–65	35–45	–	<1	–	Jönsson et al. (2003)
	57.8	38.6	0	3–7	<0.1	Spiegel and Preston (2003)
Pig manure	55–65	35–45	–	0–3	–	Polprasert and Koottatep (2007)
	42–59	–	–	–	–	Tran et al. (2014)

5.1.2 *Anaerobic Digestion Process*

AD is a biological process that degrades organic material by the concerted actions of a wide variety of microbial communities in the absence of oxygen. In a simplified description, AD is divided into four phases (Fig. 5.1): hydrolysis, acidogenesis (acid-producing), acetogenesis (acetic acid-producing), and methanogenesis (methane-producing).

5.1.2.1 Stage 1: Hydrolysis

Insoluble organic compounds, such as cellulose, protein, fat, and some insoluble forms of organic compounds, are decomposed by enzymes (produced by bacteria) and anaerobic bacteria. Small soluble organic molecules, which are produced in this stage, are the raw material for the bacteria in the next stage. Hydrolysis of carbohydrates can occur within a few hours, while protein and fat hydrolysis may take several days. However, lignocellulose and lignin substances decompose slowly and incompletely (Deublein and Steinhauser 2011). Facultative anaerobes consume dissolved oxygen in water, which results in reduction of redox potential that is favorable to the AD process. In this stage, carbohydrates are broken down into simple

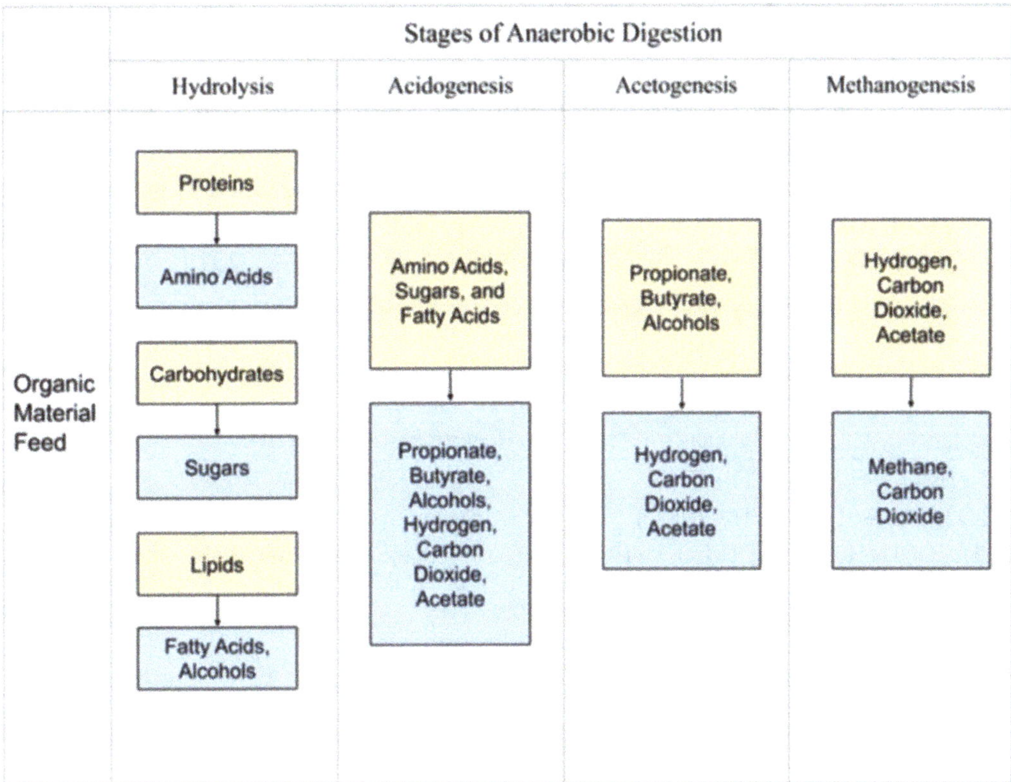

Fig. 5.1 Four stages of methane production. (Adapted from Zinder (1993))

sugars; fats degrade into fatty acids; and proteins degrade into amino acids (Gerardi 2003; Eastman and Ferguson 1981).

5.1.2.2 Stage 2: Acid-Producing (Acidogenesis)

Simple organic compounds which were created during hydrolysis stage will be transformed into volatile fatty acids (VFAs), long chain fatty acids, propionate, and butyrate by anaerobes (Jördening and Winter 2006). The concentration of H^+ formed in this stage may affect the products of fermentation. High concentration of H^+ reduces the production of acetate. In general, during this stage, simple sugars, fatty acids, and amino acids are fermented to form organic acids and alcohol (Gerardi 2003).

5.1.2.3 Stage 3: Acetic Acid-Producing (Acetogenesis)

The products from the previous stage are the substrate for bacteria in the acetic acid-producing stage. The products from these intermediate substrates are H_2, CO_2, and acetate. Acetogenic bacteria grow together with methanogen bacteria in this stage.

5.1.2.4 Stage 4: Methane-Producing (Methanogenesis)

During this stage, methane is created under completely anaerobic conditions. This reaction is considered an exothermic reaction. Stage 4 can be divided into two methane-generating processes: reduction of CH_3COO^- and conversion of H_2 with CO_2. Acetotrophic methanogens are responsible for the reduction of acetate (CH_3COO^-) into methane, while hydrogenotrophic methanogens are responsible for converting H_2 and CO_2 into methane (Ziemiński and Frąc 2012). Some of the reactions during methanogenesis are described in Table 5.2.

Table 5.2 Some methanogenesis reactions

Reactants	Products	Organisms involved
$4H_2+HCO_3^-+H^+$	CH_4+3H_2O	Most methanogens
$4HCO_2^-+H^+ + H_2O$	$CH_4+3HCO_3^-$	Many hydrogenotrophic methanogens
$2CH_3COOH+HCO_3^-$	$2CH_3COO^-+H^+ + CH_4+H_2O$	Methanosarcina and Methanothrix
$4CH_3OH$	$3CH_4+HCO_3^-+H_2O+H^+$	Methanosarcina and other methylotrophic methanogens

Source: Zinder (1993)

Table 5.3 Factors that affect the AD process

Factors	Range	Optimum value for methane production
Temperature	<20–60 °C	35 °C
pH	6.6–7.6	7.0
Redox potential	≤150 mV	≤250 mV
Salinity	0–8%	0.84.5%
C/N ratio	20–40	20–30
Loading rate	1–4 kg VS m^3 day^{-1}	1–4 kg VS m^3 day^{-1}
Retention time	10–60 days	10–30 days

5.1.3 Factors Affecting the Anaerobic Digestion Process

The AD process is affected by different factors that include operation parameters, type of feedstock, rate of feed of feedstock, etc. Some factors are enumerated in Table 5.3 and discussed below.

5.1.3.1 Temperature

Depending on the microorganisms involved, the anaerobic reactors operate in specifuied temperature regimes:

- Psychrophilic anaerobic digestion (<20 °C)
- Mesophilic anaerobic digestion (20–45 °C)
- Thermophilic anaerobic digestion (46–60 °C)

The activity of methane-forming bacteria is strongly influenced by temperature. In general, when the temperature increases, biogas production increases. For that reason, digesters are heated in colder regions, such as Europe. But at a temperature range of 40–50 °C, biogas production will decrease because this temperature range is not suitable for both mesophilic and thermophilic bacteria. If the temperature is above 60 °C, biogas production decreases and it stops completely at 65 °C or higher. Biogas production is at its maximum rate when the temperature is kept at 35 °C (Chandra et al. 2012), while optimum temperatures for methane production range from 35 to 40 °C (Lianhua et al. 2010).

The most appropriate temperature is 30–40 °C. Low temperatures, abrupt changes in the system, or both weaken methanogen activities (Yadvika et al. 2004; Nozhevnikova et al. 1999).

Temperature is a very important variable for efficient anaerobic digestion of rice straw. Maximized methane production and economic input depend on it.

5.1.3.2 pH and Alkalinity

McCarty (1964) determined that the biological process of AD works optimally in a pH environment of 6.6–7.6. However, Yadvika et al. (2004) and Gerardi (2003) reported that the optimum pH range for AD is at 6.8–7.2. At a pH lower than 5.5, acidogenic bacteria are still active but methanogenic bacteria are inhibited. Inhibition of methanogenic bacteria lowers the methane content of the biogas; therefore, low pH conditions are avoided. During the AD process, pH will usually drop lower than 6.6 if there is excessive accumulation of fatty acids in the acidogenesis stage. Decrease in pH is caused by either overloading of substrates or because the toxins in the feedstock inhibit the activity of methanogens. In this case, substrate feed must be stopped so that acid production will stop or decrease and acetogens and methanogens will be able to degrade excess acid that was produced. Another solution is to use lime for neutralizing the acid and increase the pH to an optimum range. An increase in pH (greater than 8.0) also has an inhibitory effect on AD. At pH 9.0, the methanogenesis process completely stops (Clark and Speece 1971).

Alkalinity is a measure of the capacity of a solution to neutralize acid. Bicarbonate (HCO_3^-), carbonate (CO_3^{2-}), and hydroxide (OH^-) are the ions that are used to increase alkaline conditions in the digester. Alkalinity is also considered as the buffering capacity of a solution and is essential for controlling and maintaining a stable pH for the anaerobic digestion process. High alkalinity conditions (1500–3000 mg $CaCO_3$ L^{-1}) enhance the pH stability of the anaerobic digestion process (Gerardi 2003).

5.1.3.3 Redox Potential

The redox potential is a measure of oxidation capacity or reducing capacity. Biogas is produced effectively in an anaerobic environment where the redox potential must be less than −150 mV. Redox always reaches a negative value (less than −100 mV) under anaerobic conditions (Wiese and König 2009).

In general, the use of substrates including oxygen, nitrate and sulfate promotes oxidation that can significantly change the redox potential and cause changes in pH. Redox potential can be used to predict impending changes in pH of the digester (Wiese and König 2009).

Methane begins to form and CO_2 and H_2 are converted into CH_4 when the redox value is less than −250 mV; H_2O and H_2S are produced when the redox value is less than −150 mV (Laanbroek 1990).

5.1.3.4 Salinity

High salinity and ammonium concentrations have detrimental effects on biological processes such as anaerobic digestion (Fang et al. 2011; Chen et al. 2008; Reinhart and Townsend 1998; Kargi and Dincer 1996). High salt concentrations dehydrate bacterial cells due to osmotic pressure (Alhraishawi and Alani 2018). Salt toxicity is determined mostly by the type of cation the salt has.

Feedstock inflow to anaerobic digesters usually contains light metal ions, namely, sodium, potassium, calcium, and magnesium. These cations may also be liberated during the AD process (Chen et al. 2007). A study by Albraishawi and Alani (2018) on codigestion of food waste demonstrated that increasing salt concentrations (0, 16, 30, and 60 g NaCl L^{-1}) has a negative effect on the volume of biogas produced (45, 21, 5, and 2 ml d^{-1}). In terms of methane yield, a study by Lee et al. (2009) on anaerobic digestion of leachate from a food waste recycling facility showed that low-salt concentrations (0.5 and 2 g NaCl L^{-1}) increase methane yield, but higher salt concentrations (5 and 10 g NaCl L^{-1}) resulted in a decrease in methane yield (36 and 41% reduction).

Anwar et al. (2016) showed that methane yield inhibition in anaerobic digestion of food waste is negligible at salt concentrations of 8 g NaCl L^{-1}, but salt concentrations greater than 8 g Nacl L^{-1} resulted in a sharp decline in methane yield. The cubic regression model $y = 0.508 + 2.401x - 0.369x^2 + 0.033x^3$ was derived from this experiment to describe sodium salt inhibition, where y is the methane yield and x is the sodium salt concentration. This model predicted experimental results with a small discrepancy of 10%. A study by Ogata et al. (2016) on the effect of salt on biogas production of leachate in a waste landfill showed that methane production decreased while carbon dioxide production was unchanged at a salt concentration of 35 ms cm^{-1} (approximately 19 mg L^{-1}). A salt content of 80 ms cm^{-1} (approximately 44 mg L^{-1}) decreased production of both methane and carbon dioxide. Based on these studies on salt inhibition of the anaerobic digestion process, it can be inferred that low salt concentrations in the AD reactant mixture (up to 2 g L^{-1} NaCl) increase methane yield, but higher salt concentrations (greater than 5 g L^{-1} NaCl) decrease methane yield.

5.1.3.5 Carbon to Nitrogen (C/N) Ratio

The quality of biogas produced by anaerobic digestion is determined by the growth of the community of bacteria in the digester. The optimal carbon to nitrogen (C/N) ratio for bacteria to grow is in the range of 20–30 because the bacteria use up carbon 20 to 30 times quicker than nitrogen (Bardiya and Gaur 1997; Malik et al. 1987). If the C/N ratio is higher than optimal, the decomposition rate will be slower. When the C/N ratio is low, the accumulation of ammonia can occur, which can inhibit the activity of bacteria. Some substrates for AD with different C/N ratios are shown in Table 5.4.

The difference in C/N ratios show that plant materials have a high C/N ratio and animal manures have a low one. To achieve the optimum C/N ratio of 20/1–30/1, plant material and animal manures are codigested.

The C/N ratio is a critical factor in the anaerobic digestion process, which shows the balance of nutrients of input materials. Depending on the type of paddy rice, untreated rice straw has a low concentration of total N content, even less than 1% of

Table 5.4 C/N ratio of some AD substrates

Organic source	C/N ratio	Source
Rice straw	44.0–74.2	Li et al. (2015), Gu et al. (2014), Ye et al. (2013); Lim et al. (2012), Hills (1981)
Water hyacinth	12–42	Ngan et al. (2011)
Cow manure	13.0–14.2	Biosantech et al. (2013), Hills (1981)
Pig manure	7.0; 10.8	Biosantech et al. (2013), Hills (1981)
Chicken manure	4.4; 7.0	Biosantech et al. (2013), Hills (1981)

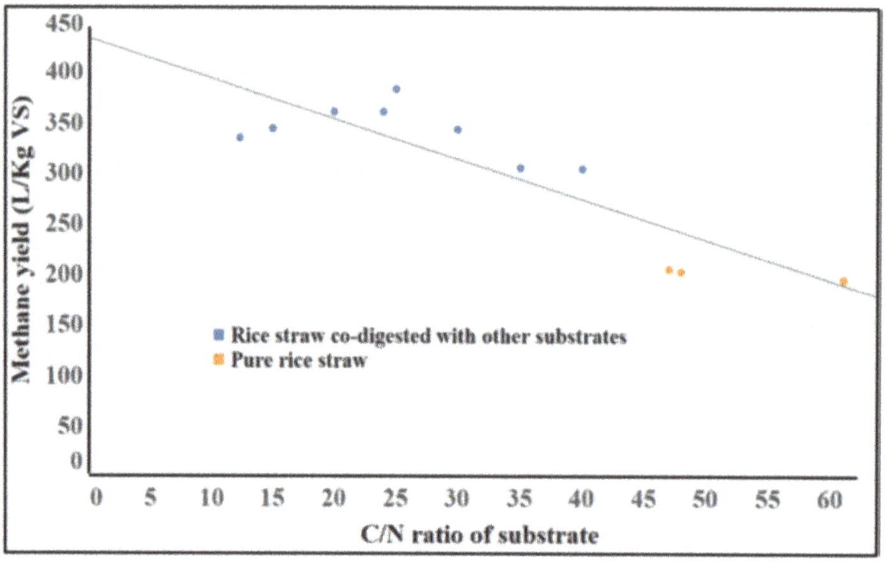

Fig. 5.2 Effect of the C/N ratio on methane yield of rice straw. (Sources: Yan et al. (2015), Ye et al. (2013), Menardo et al. (2012), Dinuccio et al. (2010), and Hills (1981))

dry basis (Dinuccio et al. 2010; Lei et al. 2010). The typical C/N ratios for untreated rice straw are approximately from 40 to 80 (Tran et al. 2014; Arvanitoyannis and Tsherkezou 2008; Ghosh and Bhattacharyya 1999). The effect of the C/N ratio on methane yield is illustrated in Fig. 5.2, which demonstrates that methane yield increases when C/N ratio reaches the optimum range (20/1–30/1) and decreases when the C/N ratio rises beyond the optimum.

5.1.3.6 Loading Rate and Hydraulic Retention Time (HRT)

The AD process is affected by the loading rate, represented by chemical oxygen demand per cubic meter per day (COD m^{-3} day^{-1}) or volatile solids per cubic meter per day (VS m^{-3} day^{-1}), and hydraulic retention time (HRT) of the fermentation mixture. A high organic loading rate will cause accumulation of fatty acids at stage

3, which negatively affects the production of methane. A low organic loading rate on the other hand will produce low volumes of biogas that will make AD operation uneconomical. The recommended organic loading rate for an anaerobic tank without medium is 1–4 kg V m^{-3} day^{-1} (Eder et al. 2006). HRT is the time that the substrates stay inside the anaerobic digester reactor. HRT strongly depends on the substrates and it usually varies from 30 to 60 days. In general, 30 days are typical HRT for nonstirring digesters, the digesters with high decomposition rates can be reduced to an HRT of 10–20 days.

5.1.3.7 Toxins

Biogas production happens in the absence of oxygen; therefore, the presence of oxygen will inhibit the process. In this case, oxygen is considered to be toxic to anaerobic bacteria. In addition, there are many substances that are potentially toxic to microorganisms. Toxic substances and their toxic doses are presented in Table 5.5.

5.1.3.8 Dry Matter and Water Content

The percentage of dry matter content in a digester suitable for biogas generation and solids reduction is about 9–10%. The concentration of dried matter up to 20% helps to save 50% of the volume of digesters but may lead to souring (reduction of pH) and consequently, reduction of biogas output. Monnet (2003) recommended that the organic dry matter content of batch digester should be adjusted to a range of from 5% to 10%.

Table 5.5 Toxic dose of substances on AD

Substances	Toxic dose for AD bacteria	Sources
Volatile fatty acids	>10,000 mg L^{-1}	Wang et al. (2009)
Oleate	>1700 mg L^{-1}	Angelidaki and Ahring (1992)
Stearate	>1000 mg L^{-1}	Angelidaki and Ahring (1992)
NH_3	>16,000 mg L^{-1}	Koster and Lettinga (1988)
S_2^-	>145 mg L^{-1}	Parkin et al. (1990)
Ca	>7000 mg L^{-1}	Ahn et al. (2006)
Mg	>1000 mg L^{-1}	Gerardi (2003)
K	>8000 mg L^{-1}	Kugelman and McCarty (1965)
Na	>3500 mg L^{-1}	Gerardi (2003)
Fe	>5 mg L^{-1}	Gerardi (2003)

5.1.3.9 Stirring

Stirring maximizes contact of bacteria with organic waste to accelerate the digestion process. It also minimizes the solids deposition on the bottom of the digesters and avoids foaming and scum on the fermentation solution surface.

Without stirring, the substrate in the digester is usually stratified into three layers: the upper layer is the floating layer; the middle layer is fermentation solution; the bottom layer is the sediment layer (Fig. 5.3). Bacteria not distributed evenly in fermentation solution result in uneven contact of bacteria with the substrates. There are many "dead zones" in the digester where the bacteria density is very low and the decomposition is weak. Organic materials can accumulate and settle in those zones. Stirring can overcome the above disadvantages and enhance the decomposition process.

During the digestion process, it is necessary to mix the substrates in the digester, especially for the substrate from plant biomass, to avoid scum formation. Mixing increases the contact between bacteria and substrate and improves decomposition. Mixing can also prevent foaming and make the temperature uniform inside the digester. However, a fast mixing speed will break down the microbial population, thus it is best to stir at lower mixing speeds (Monnet 2003).

5.1.3.10 Feedstock Pretreatment

Fibrous substrates—especially straw, grass, weeds, and stalks—are difficult to decompose and must be treated before digestion. Pretreatment aims to reduce the crystallization of cellulose, increase the surface area of the substrate, and make cellulose more accessible to enzymes that convert carbohydrates into fermentable

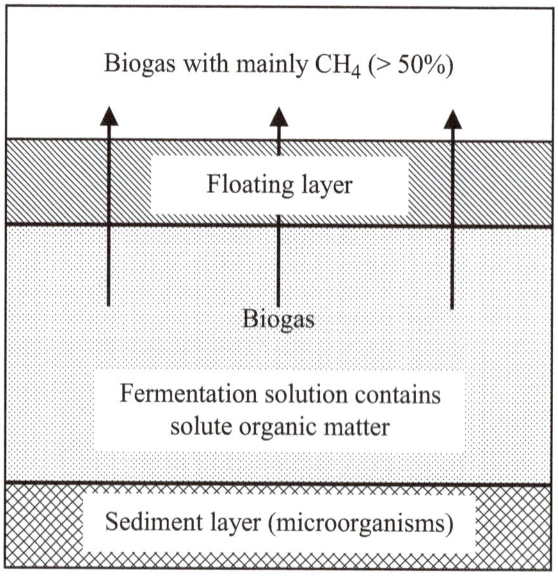

Fig. 5.3 Layers in an unstirred anaerobic digester

sugars. Pretreatment includes physical, chemical, biological, and their combinations (Alvira et al. 2010).

The anaerobic fermentation process can be enhanced if the digestate is recycled and mixed with substrate feed. The solids content of feeding materials should be adjusted to 5–10%.

5.1.3.11 Feedstock Size

Feeding material size is one of the factors affecting biogas production. Materials should not be too large because it will lead to digester blockage and will also make it difficult for bacteria to decompose. Small-sized materials will have larger surface areas and increased microbial activity, resulting in faster decomposition. According to Sharma et al. (1988),when substrates having five different particle sizes (0.088, 0.4, 1.0, 6.0, and 30.0 mm) were digested and their biogas output were compared, it was shown that the optimal size for producing biogas was 0.088 and 0.4 mm. Some other studies also suggested that a physical pretreatment method, such as grinding and chopping, can significantly reduce digester size design as compared with digesters using untreated substrates,without reducing the biogas production (Moorhead and Nordstedt 1993; Gollakota and Meher 1988).

5.2 AD Systems

5.2.1 Small-Scale Biogas Digesters

Small-scale biogas plants have mostly been popularized in developing countries. In many of those countries, massive government-led (China, India) or government-NGO biogas programs were initiated with the aim of popularizing this technology.

Based on the construction methodology, there are two main types of plants including:

- Constructed on-site plants: these plants are often made of brick, mortar and concrete.
- Prefabricated plants (PBD): these plants are produced off-site and installed at the farms. They are made of fiber-reinforced plastic (FRP), tubular (known as bag plants or soft plastic plants), and hard plastics, such as hard polyvinyl chloride (PVC), acrylonitrile butadiene styrene (ABS), high-density polyethylene (HDPE), linear low-density polyethylene (LLDPE), and modified plastics.

5.2.1.1 Biogas Plants Constructed On-Site

These plants include widely developed and used technologies such as fixed-dome, Indian, and floating drum digester types shown in Fig. 5.4a–c, respectively. Features of these digester types are described in Table 5.6.

5.2.1.1.1 Fixed-Dome Digesters

The archetype of all the on-site constructed hydraulic biodigesters is the water pressure digester of Luo Guorui developed in the 1920s in Taiwan (Nianguo 1984). This design eventually led to the Chinese fixed-dome digester. The gas pressure varies

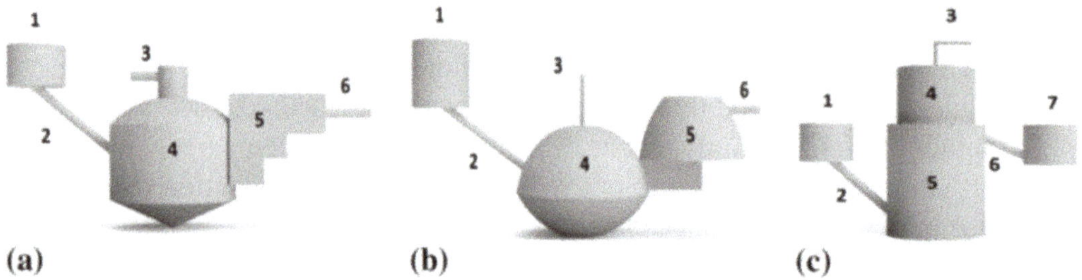

Fig. 5.4 (**a**) Chinese-style fixed dome digester, (**b**) Indian-style fixed dome digester, (**c**) Floating drum digester. 1 = Feedstock mixing tank, 2 = Inlet pipe, 3 = Biogas outlet, 4 = Digester, 5 = Displacement chamber, 6 = Slurry outlet pipe, 7 = Sludge collection tank

Table 5.6 Comparison of digester plants constructed on-site

Digester type	Advantages	Disadvantages
Top manhole (Chinese plant types)	Easy to enter the digester from the top	Top manhole is situated on top of gasholder and is prone to leakages
	Can withstand high pressures, therefore, can store more gas	Gas proofing of the dome requires special paints and skilled masons
Manhole at the outlet (Deenbandhu, Janata)	Has the highest load bearing capacity as the dome is a closed solid structure	Gas proofing of the dome requires special paints and skilled masons
	Cheaper than Chinese digester models due to optimized shape	
Floating drum	Drum is gastight	Higher investment and additional transport cost of drum
	Lower masonry skill requirement	Distance between inlet and outlet is relatively short which can result in preferential flow of manure shortening the HRT
		High depth to width ratio which makes construction difficult in certain soils

depending on the amount of slurry accumulated in the displacement tank (or compensation / hydraulic chamber). The pressure is at its highest when the gasholder is full of gas when an equivalent volume of slurry is push upward into the displacement tank; and the pressure is 0 when all gas is used. Gas pressure typically varies between 2 and 15 kPa (Gunnerson et al. 1986).

5.2.1.1.2 Indian Digesters

The Deenbandhu digester is the archetype of all fixed-dome digesters with a manhole at the outlet tank. The digester can resist higher structural forces compared to the Chinese digester types as the dome is a closed structure. Another advantage is that the gasholder is closed that reduces the risk of leaks compared to a Chinese digester, which has a manhole at the top of the gasholder (Balasubramaniyam et al. 2008).

5.2.1.1.3 Floating Drum Digesters

The floating drum digester was introduced by the Khadi and Village Industry Commission and was later branded as the KVIC digester (Singh and Sooch 2004). The floating drum digester is fed semicontinuously and has a relatively high depth-to-width ratio. A barrier is placed in the middle of the digester to promote mixing and prevent short-circuiting (direct substrate flow from the entrance to the exit) (Gunnerson et al. 1986). Biogas accumulates at the headspace of a movable inverted steel drum that serves as the upper part of the digester.

5.2.1.2 Prefabricated On-Site Biogas Plants

5.2.1.2.1 Fiber-Reinforced Plastic (FRP) Digesters

The technical designs of fixed-dome FRP biodigesters can be classified in different ways according to the major features of the digester, such as the full FRP digester and the partial FRP digester (usually upper-dome part only) based on the integrity of FRP structural materials; top-mounted hydraulic chamber and lateral-placed hydraulic chamber based on the placement of hydraulic (displacement) chamber; spherical, ellipsoidal, and cylindrical based on the shape of digester body; modular type and integral type based on assembly work; and differences in digester volume. Sizes of these plants are typically in the range of 4–9 m^3.

5.2.1.2.2 Hard-Plastic Digester Types

The most common processes used to produce hard-plastic biodigesters are rotation and injection molding. Injection molding machines are more expensive, but the process allows for more intricate shapes, can utilize more types of materials, and is much faster, resulting in lower production costs. Materials used include hard PVC, ABS, HDPE, LLDPE, and modified plastics.

5.2.1.2.3 Soft-Plastic Digester Types

The archetype of soft-plastic digester types is the Taiwanese bag digester, which was developed in the 1960s. It is a popular model especially in Central and South America (Gunnerson et al. 1986). The digester is made of a cylindrical plastic tank (tube) nested horizontally on a hardened layer of masonry, concrete, sand, or mud (Gunnerson et al. 1986). A great advantage of the bag digester is the simple design and low material costs. However, these low-cost materials often break down easily and often render the effective lifespan of the digester to less than 2 years. In the last decade, SP digester designs emerged that use higher-quality materials, such as geomembranes and HDPE instead of polyethylene. Such digesters are durable and should have a lifespan of from 6 to 10 years. Lower-quality soft-plastic digesters made from simple PE sheets are also available. These digesters are cheap, but are easily damaged and require frequent repairing and, as such, are not actively promoted.

5.2.2 *Medium- and Large-Scale Plants*

Two-stage AD is widely used for medium- and large-scale biogas plants with the two digester tanks in series. The advantage of having two stages for AD is the separation of the methanogenesis stage from the acidogenesis stage because acids produced in the acidogenesis stage can inhibit the methanogens. The first tank is optimized for hydrolysis and acidogenesis and the second tank is optimized for acetogenesis and methanogenesis. An example of the two-stage AD power plant is presented in Sect. 5.3.2.2.

Medium- and large-scale biogas plants are used on commercial farms and often produce more biogas than the farming households can use. Consequently, biogas is turned into electricity or supplied to gas grids or fuel stations after being upgraded to a desired purity. Most of the large-scale biogas plants are constructed in Germany, USA, UK, and China. Many of the these plants are generating biogas from landfill, wastewater, fuel crops, and to a lesser extent from livestock manure.

5.3 Current Technology Developments and Practices for Rice Straw AD

5.3.1 *Rice Straw Pretreatment for AD*

Pretreatment methods to improve the anaerobic digestion process were presented in many studies, such as Ngan (2012), Pilli et al. (2011), Hendriks and Zeeman (2009), Neyens et al. (2003), Weemaes and Verstraete (1998), Stuckey and McCarty (1984), and Haug et al. (1978). Pretreatment has been reported as an important step in the methane production process (Alvira et al. 2010; Carvalheiro et al. 2008; Taherzadeh and Karimi 2008). Pretreatment will change the structure of cellulose so that the enzymes can easily convert high molecular weight molecules, such as carbohydrates into simple sugars (Mosier et al. 2005). Especially in the case of rice straw, a biomass with a high-lignin pretreatment step is necessary to amplify the degradability of rice straw and speed up the anaerobic digestion process.

Rice straw pretreatment methods are classified as physical (particle size reduction), chemical (acid and alkali additions), and biological (fungi).

5.3.1.1 Physical Pretreatment: Effect of Particle Size of Rice Straw

Pretreatment of the feedstock can increase its solubility, consequently increasing biogas production and enhancing reduction of volatiles and solids content. Pretreatment is especially helpful in the digestion of biomass substrates as these substances tend to have high cellulose or lignin content. Additives can increase the production rate of the reactor or increase the startup speed, but their additional cost must always be balanced against improvements in efficiency (Ward et al. 2008; Ngan 2012).

Rice straw particle size reduction breaks the cell walls and makes the organic substrate more readily available for microbes to decompose (Zhang and Zhang 1999). Size reduction of rice straw increases surface area and breaks down its polymer structure, thereby increasing hydrolysis yield and hydrolysis rate during digestion (Hendriks and Zeeman 2009). Gharpuray et al. (1983) verified that pretreatment of wheat straw by ball-milling was found to be effective in increasing specific surface area (2.3 $m^2\ g^{-1}$ for pretreated substrate compared to 0.64 $m^2\ g^{-1}$ for raw straw). Fiber degradation and methane yield are enhanced when particle size is reduced from 100 mm to 2 (Mshandete et al. 2006). Consequently, methane yield increased (5–25%) and digestion time was reduced (23–59%) (Hendriks and Zeeman 2009). A study by Zhang and Zhang (1999) revealed that rice straw cut in 25-mm lengths has higher methane (198 L kg^{-1} VS) versus uncut rice straw. However, addition of a milling step in the AD process is expensive due to its high energy requirements (Hendriks and Zeeman 2009).

Møller et al. (2004) observed, for rice straw AD, the increase in methane yield from 30-mm lengths (145 L kg^{-1} VS added) compared to 1-mm lengths (161 L kg^{-1} VS added) was significant after 60 days of digestion. Increased bio-degradability of rice straw has been demonstrated when methane production yield was found 7.9 and 13% higher, respectively, for sizes of 0.30 and 0.75 mm compared to 1.5 mm (Chandra et al. 2015). Sharma et al. (1988) reported that methane production tends to increase by from 43.5 to 52.0% when the particle size was gradually reduced from 30 mm to 0.088 mm. However, rice straw sizes under 0.5 mm could be detrimental to the methane production process due to excessive accumulation of volatile fatty acids (VFAs) that will reduce pH in biogas reactor, causing inhibition of anaerobic organisms.

Methane production from rice straw with sizes of 2, 5, and 10 mm illustrates that there is no significant difference among treatments, but compared to the untreated straw, methane production is from 8.3 to 9.3% higher. Comparing the methane yields of AD of rice straw of different sizes (1, 10, 20 cm, and original size) that was soaked in AD slurry for 5 days, it was found that there is no substantial difference in methane yields (166, 121, 166, and 168 L kg^{-1} VS added, respectively) between treatments (Fig. 5.5).

Chopping and grinding rice straw into small particles less than 1 mm in size and mixing with kitchen waste and pig manure could produce methane from 205 to 384 L kg^{-1} VS added (Ye et al. 2013) whereas 2–3 cm of chopped rice straw with added micronutrients (nickel and cobalt at 10 and 15 mg kg^{-1}, respectively) increased biogas production by 37 and 46% in mesophilic and thermophilic digesters, respectively.

5.3.1.2 Chemical Pretreatment

Chemical pretreatment methods mainly include acid and/or alkaline pretreatments.

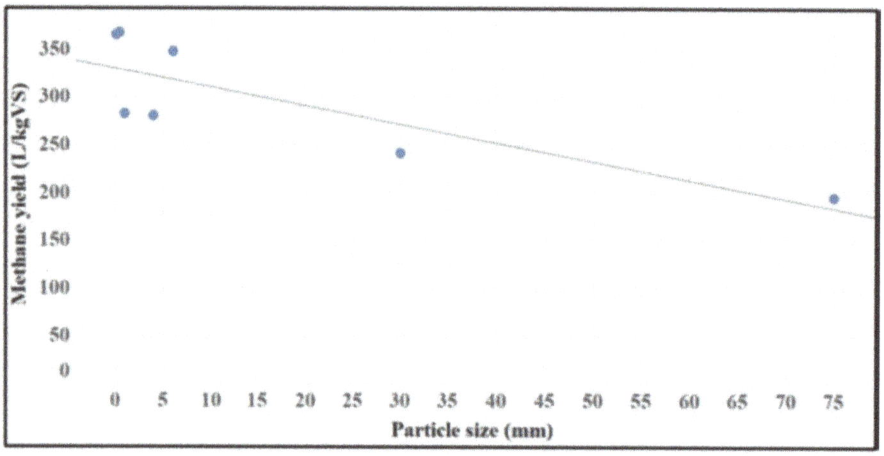

Fig. 5.5 Methane yields of rice straw at different particle sizes. Data sources: Ye et al. (2013), Menardo et al. (2012), Lei et al. (2010), Dinuccio et al. (2010), and Sharma et al. (1988)

5.3.1.2.1 Acid Pretreatment

Acid pretreatment is desirable for anaerobic digestion of rice straw because it breaks down the lignin structure and helps methanogens acclimatize to low pH conditions.

- Hydrolysis by weak acid: acid pretreatment at low concentration is one of the most effective methods for treating lignocellulosic biomass. This method includes two types of hydrolysis: (1) continuous hydrolysis at high temperature (above 160 °C) and a low loading rate (about 5–10% TS), and (2) batch hydrolysis at low temperatures (below 160 °C) with a high loading rate (about 10–40% TS). Acid sulfuric—sprayed into the lignocellulose, mixed, and then kept at from 160 to 200 °C for a few minutes—will remove hemicellulose. As a result, the efficiency of the hydrolysis process will be improved. Acid pretreatment showed the significant improvement of enzyme activity and removed hemicellulose effectively (Chen et al. 2007).
- Strong acid pretreatment: High concentrations of H_2SO_4 and HCl are widely used for pretreatment of lignocellulose because these acids are powerful and they help hydrolyze cellulose (Sun and Cheng 2002) without the involvement of enzymes in the hydrolysis process. The disadvantage of this method is the corrosive properties of these acids and they should be recycled to reduce pretreatment costs.

5.3.1.2.2 Alkaline Pretreatment

The main effect of alkaline pretreatment is lignin removal from rice straw, therefore improving the degradability of polysaccharides. In addition, alkaline pretreatment also eliminates acetyl and replaces uronic acid on hemicellulose, thereby increasing the enzyme activities on hemicellulose and cellulose surfaces.

- Pretreatment with ammonia solution: This is an effective treatment for lignocellulose. An ammonia solution is highly selective in reactions with lignin in comparison to carbohydrates. One of the reactions in an ammonia solution with lignin is the breakdown of the C-O-lignin as well as the ether and ester bonds in the lignin complex (Binod et al. 2010). However, this solution is an environmentally-polluting compound and a corrosive chemical.
- Pretreatment with calcium and sodium hydroxide: Lime ($Ca(OH)_2$) and sodium hydroxide (NaOH) are commonly used for the pretreatment process. During the process. Salts can be formed and incorporated into materials (González et al. 1986). The condition of this process is quite simple but the reaction time is relatively long. This pretreatment results in high solubility of lignin, especially for materials with low lignin content such as softwoods and weeds. The addition of air or oxygen during the pretreatment process can improve the efficiency of the lignin decomposition (Chang and Holtzapple 2000).

5.3.1.3 Biological Pretreatment

Biological pretreatment of feedstock for AD has attracted interest because it requires less energy input as compared with physical pretreatment and is less costly than chemical pretreatment, which costs more because of the expensive chemicals required. Hence, the biological route seems to be the most promising because it is an eco-friendly process and there is no inhibition during the process (Liu et al. 2014).

Rice straw is one of the most important biomass energy sources, largely because of its abundance. According to the Vietnam Statistical Yearbook (2018), around 23.63 million tons of rice straw are produced annually in the Mekong Delta, but more than 80% of it is burned on-site (Nguyen and Tran 2015). Typically, rice straw has a complex polymer crystal structure that is formed by the physical and chemical bonds among the cellulose, hemicellulose, and lignin components, which renders it difficult for anaerobic bacteria to utilize these components for biogas production (Sun et al. 2015). This becomes a major limitation to rice straw's efficient utilization.

Generally, methane yields from agricultural biomass are lower compared to conventional substrates, but agricultural biomass is an inexpensive option. Biological pretreatment methods can reduce anaerobic digestion duration, enhance feedstock digestibility, and increase gas production rate. This is because the lignocellulosic components of the straw are degraded into simple substances and made easy to digest in AD, especially when using microorganisms with strong lignocellulose degradation ability. The key to the success of biological pretreatment is to find microorganisms that have exceptional lignin degradation ability and to determine the optimum digestion conditions for these microorganisms. Examples of biological pretreatment methods are: microaerobic treatments, ensiling or composting, separation of digestion stages, and fungi pretreatments.

Biological pretreatment of rice straw using fungi is a comparatively eco-friendly approach of enhancing degradability when compared with chemical pretreatment, which requires expensive chemicals, high energy inputs, and toxic substance removal, (Carrere et al. 2016). Several fungi species are used for pretreatment of lignocellulosic biomass for anaerobic digestion and most of them are the white-rot fungus (*Ceriporiopsis subvermispora*). A study by Zhao et al. (2014a, b) using white-rot fungus as the pretreatment agent increased methane yield by 5–15% as compared with untreated biomass.

Biological pretreatment is normally done by soaking the straw in a natural microbial solution obtained from the effluent of an anaerobic digester, anoxic sediment from ponds or lakes, and wastewater. One study found that rice straw pretreated by soaking in anoxic sediment and digester effluent for 5 days produced 79–85% more biogas volume than straw soaked in tap water (Tran et al. 2017). Compared with untreated substrates, pretreatment using microbiological action increases the degradability of substrates. Yadav et al. (2019) verified that optimal conditions for the biological treatment of lignocellulose biomass of wheat straw by *Chaetomium globosporum* was found to be 36 °C, 31 days, and 81% moisture, resulting in a 2.9-fold increase in reducing sugar, 48% removal of lignin, and 31% increase in biogas yield.

Similarly, a study by Shen et al. (2018) on the effect of organic loading rate on anaerobic codigestion of rice straw and pig manure showed that after biological pretreatment, the substrate was optimally fermented at an organic loading rate of 2.5 kg COD m^{-3} day^{-1}. This pretreatment achieved the optimum volumetric methane production rate of 640 L CH_4 m^{-3} day^{-1}, and a methane yield of 456 L CH_4 kg^{-1} COD removed, which were 62.4 and 37.8% higher than those of the control under the same organic loading rate.

Using aerobic and anaerobic fungi as pretreatment agents resulted in increased biodegradability of rice straw (Ghosh and Bhattacharyya 1999; Cann et al. 1994). Methane production from rice straw AD was increased 31–46% when pretreating it with white-rot fungus and brown-rot fungus (*Polyporus ostreiformis* using a straw-to-fungi ratio of 14:1 and then digested in batch reactors at 30 °C after a 3-week incubation period. Cann et al. (1994) reported that the anaerobic fungi from the rumen consistently increased the digestibility of rice straw when compared with fermenters where the fungi were inhibited. Haruta et al. (2002) also presented the enhancements by a microbial community formed by mixing rice straw, chicken feces, pig feces, cattle feces, and sugarcane dregs. This degraded 60% of the rice straw within 4 days.

Momayez et al. (2018) presented an investigation using effluent of biogas digestate to pretreat rice straw. The straw was pretreated at different temperatures (130, 60, and 190 °C) at different pretreatment times (30 and 60 min). The pretreated straw was subjected to different processes, including liquid anaerobic digestion (L-AD) and dry anaerobic digestion (D-AD). The highest methane yields were obtained through L-AD and D-AD of straw pretreated at 190 °C and 30 min, resulting in 24 and 26% increases in methane produced as compared with L-AD and D-AD using untreated straw. Mustafa et al. (2016) showed that rice straw pretreated with fungi (*Pleurotus ostreatus* and *Trichoderma reesei*) and used as feedstock for AD improved the methane yield by 120 and 78.3%, respectively, as compared with untreated straw. These points of view show that biological pretreatment studies have a huge potential, considering that there were only a few fungi species that were studied. There are still millions of fungi species that are yet to be studied.

The results of the previous research mentioned above are shown in Fig. 5.6.

5.3.2 *Current Practices of Rice Straw AD*

Continuous farm-scaled AD with a plastic digester was assessed and reported by Tran et al. (2015). Figure 5.7a, b show the schematic diagram of the rice straw AD system using a biogas bag developed by Can Tho University in Vietnam. The 6-m^3 household digester is made of high-density polyethylene (HDPE). Rice straw was cut in 20-cm lengths and ensilaged about 5 days before feeding into the digester. The cofeeding of rice straw and pig dung with the mixed ratio was 1:1 based on the organic dry matter (ODM). The daily feed of rice straw was 4.75 kg dry weight (DW) corresponding to 4.17 kg DW of pig dung. Retention time of rice straw in this

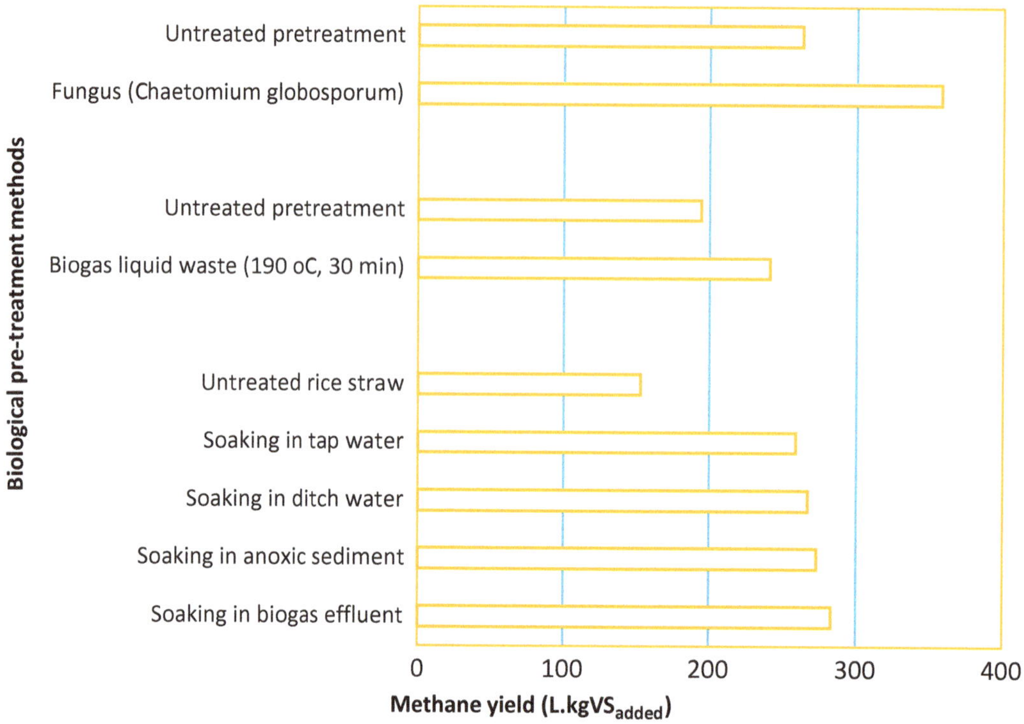

Fig. 5.6 Biological pretreatment methods of rice straw for biogas production

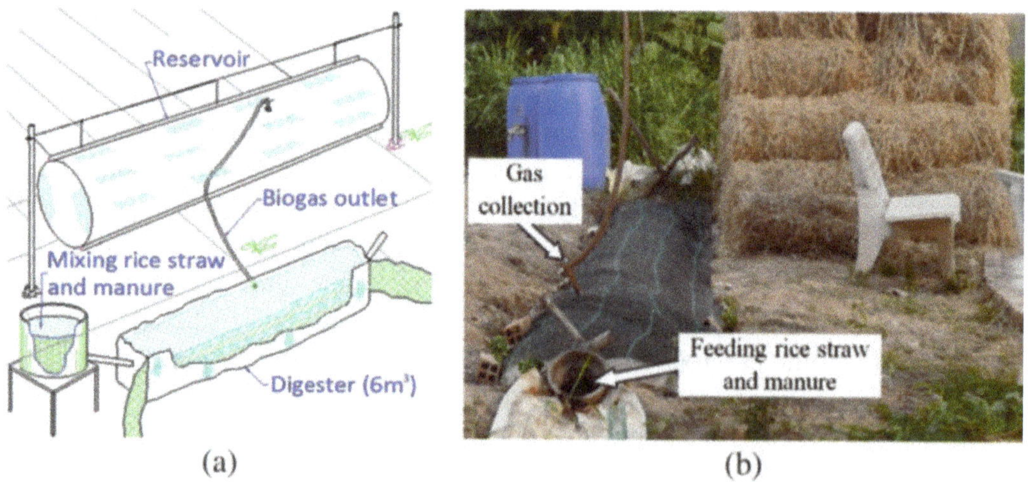

Fig. 5.7 Continuous farm-scale AD: (**a**) isometric cut-away view; (**b**) as practices in Vietnam

AD was from 110 to 120 days. Average biogas yield was 600 L kg^{-1} ODM fermented. Some constraints of this technology are the pH value dropping due to the cumulative total of volatile fatty acids (TVFAs), short-retention time due to accumulated rice straw inside the digester, and limited skill of management that may cause biogas leaking and environmental harm.

In current models, the rice straw floats on the surface of the substrate and gets stuck. Scientists at Can Tho University are testing a new HDPE model, which could prevent the floating straw by adjusting the inlet and outlet balance level. This upgrade could also improve the short retention time of the straw as well.

5.3.2.1 Rice Straw Batch AD

This technology was developed at IRRI (RKB, accessed 2019) by Nguyen et al. (2016) using a hermetic bag (so called IRRI super bag) to make the digesters. This plastic batch-AD is shown in Fig. 5.8a, b. Rice straw and carabao dung are arranged in layers in the digester. Rice straw is spread on the first layer then covered by a dung layer. This is repeated with cattle dung on the top to cover all substrates.

Biogas yield was in the range of 211–779 L kg^{-1} ODM. This experiment illustrated the following advantages of the IRRI super-bag AD: (1) the capital requirement is low with bags costing US$ 3 each, (2) floating rice straw is avoided; (3) it is portable; and (4) the digester contents are easy to unload after digestion is finished.

5.3.2.2 Two-Stage AD

Based on the assessment and verification of AD industrial technology, the pilot of a two-stage AD system was designed and is being tested at IRRI. Figure 5.9 shows the schematic diagram of the system with the following characteristics:

- Two-stage AD;
- Digestion temperature is maintained at from 35 to 55 °C;
- Feedstock: chopped or sheared rice straw mixed with animal manure based on the ratio of 75 and 25% of organic dry matter, respectively;
- Outputs: biogas for generating heat or power and digestate to produce solid and liquid fertilizer.

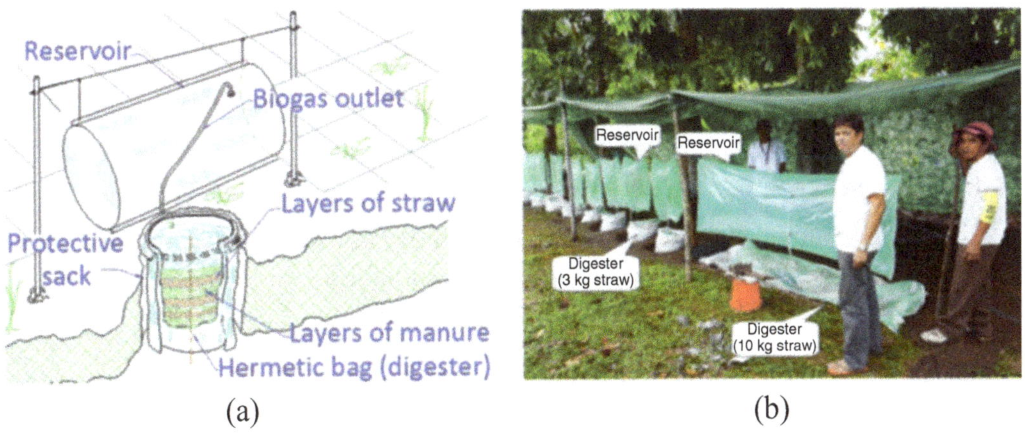

Fig. 5.8 Plastic batch-AD: (**a**) isometric cut-away view, (**b**) as practiced at IRRI

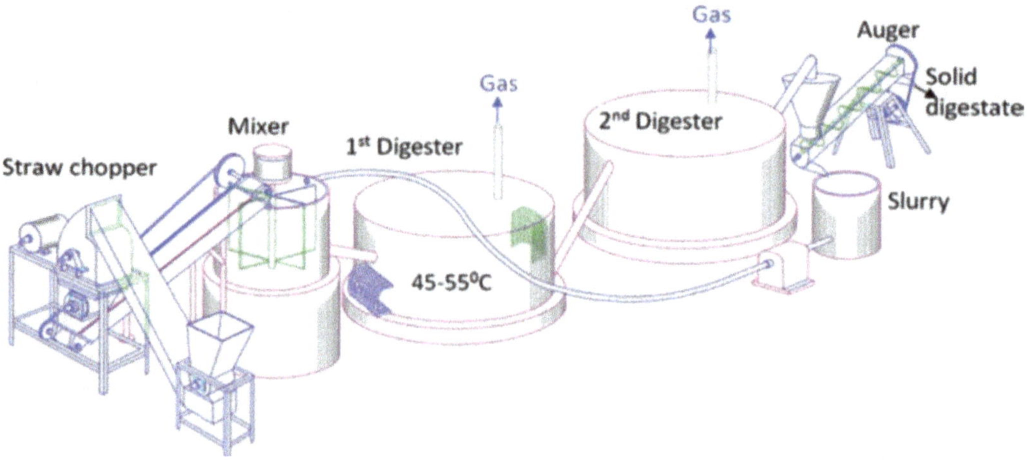

Fig. 5.9 Schematic diagram of the two-stage AD

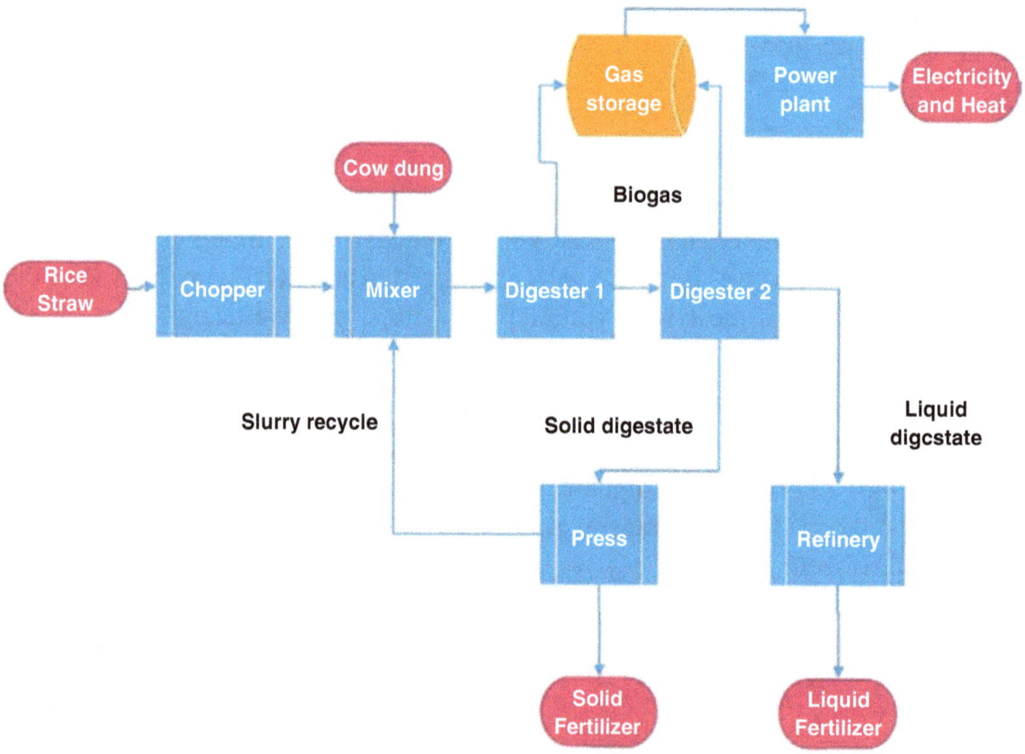

Fig. 5.10 Process route of AD power plant in Fazilka, Punjab, India

Figure 5.10 shows the process route of the biogas power plant using rice straw feedstock, which is located at Fazilka in Punjab, India. Rice straw is fed into a chopper to be sheared off. The sheared rice straw is mixed with 25% cow dung in weight and pumped into the first stage 2000-m^3 digester, which is maintained at 35 °C. Retention time (RT) of the substrate in the first anaerobic digestion (AD) stage is 20 days and then conveyed into the second-stage digester, also 2000 m^3 and maintained at 35 °C with an RT of 20 days. The biogas is collected through a reservoir and then converted to electric power by a generator. The digestate after AD is used for processing organic fertilizers.

5.4 Utilization of Bioslurry

Applications of bioslurry on crops have been investigated since the 1940s. Gurung (1997) showed that bioslurry had a better effect on crops compared to farmyard manure. This study also revealed that AD converts 25–30% of the organic part of fecal matter into biogas, while 70–75% goes to the effluent or bioslurry.

Nutrients (N-P-K) and micronutrients (zinc, iron, manganese, and copper) in bioslurry are needed for plant growth (Tripathi 1993 as cited by Gurung 1997). Bunyeth and Preston (2004) and Sophea and Preston (2001) reported that water spinach yield responded linearly to increasing levels of nitrogen in bioslurry with pig manure. The yield of leaf mustard fertilized by the codigester effluent was 2.2 times higher compared to the inorganic fertilizer treatment. In addition to increasing the yield, the effluent can help retain more nutrients in the soil layer, accelerate flower formation, and shorten cultivation time (Nguyen et al. 2015).

Using bioslurry for fisheries was also reported in several studies. Kaur et al. (1987) presented that the growth rate of carp was 3.5 times higher in the bioslurry tank than in the control tank with raw cow dung. Increases in fish growth rate using biogas effluent were also revealed by Balasubramanian and Bai (1994) and Sophin and Preston (2001).

5.5 Conclusions and Recommendations

This review of the AD process, including AD systems from different countries, provides an overview of the current practices and trends. It should be noted that there is no such thing as a best digester design as the technology is often highly localized. An AD technology may work well in one country but not in another because of different conditions.

With regards to the use of rice straw for AD, studies have shown that it is a feasible and sustainable technology, especially when rice straw is codigested with other biological wastes such as animal manure. Using the optimum pretreatment, conditions, and operational parameters and a mixture proportion with other substrates, methane production from AD of rice straw can be maximized. In addition, the digestate byproduct can be processed to produce biofertilizer.

To increase the adoption of AD technologies, awareness of the technology must be increased and the rice straw and biogas value chain must be upgraded. AD technologies that are easy to adopt, such as the use of hermetic bags for AD, can be useful to farmers. Subsequent increase in demand for biogas and biofertilizers can be supported by larger scale technologies such as the two-stage AD.

References

Ahn JH, Do TH, Kim S, Hwang S (2006) The effect of calcium on the anaerobic digestion treating swine wastewater. Biochem Eng J 30(1):33–38

Alhraishawi A, Alani W (2018) The cofermentation of organic substrates: A review performance of biogas production under different salt content. Paper presented at the sixth scientific conference, renewable energy and its applications, pp 1–14

Allen MR, Braithwaite A, Hills CC (1997) Trace organic compounds in landfill gas at seven UK waste disposal sites. Environ Sci Technol 31(4):1054–1061

Alvira P, Tomas-Pejo E, Ballesteros M, Negro MJ (2010) Pretreatment technologies for an efficient bioethanol production process based on enzymatic hydrolysis: a review. Bioresour Technol 101(13):4851–4861

Angelidaki I, Ahring B (1992) Effects of free long-chain fatty acids on thermophilic anaerobic digestion. Appl Microbiol Biotechnol 37(6):808–812

Anwar N, Wang W, Zhang J, Li Y, Chen C, Liu G, Zhang R (2016) Effect of sodium salt on anaerobic digestion of kitchen waste. Water Sci Technol 73(8):1865–1871

Arvanitoyannis IS, Tserkezou P (2008) Corn and rice waste: a comparative and critical presentation of methods and current and potential uses of treated waste. Int J Food Sci Technol 43(6):958–988

Balasubramanian P, Bai RK (1994) Biogas-plant effluent as an organic fertiliser in fish polyculture. Bioresour Technol 50(3):189–192

Balasubramaniyam U, Zisengwe LS, Meriggi N, Buysman E (2008) Biogas production in climates with long cold winters. WECF, Wageningen

Bardiya N, Gaur A (1997) Effects of carbon and nitrogen ratio on rice straw biomethanation. J Rural Energy 4(1–4):1–16

Binod P, Sindhu R, Singhania RR, Vikram S, Devi L, Nagalakshmi S, Kurien N, Sukumaran RK, PandeyA (2010) Bioethanol production from rice straw: an overview. Bioresour Technol, 101(13):4767–4774

Biosantech TAS, Rutz D, Janssen R, Drosg B (2013) Biomass resources for biogas production. The Biogas Handbook, In, pp 19–51

Bunyeth H, Preston T (2004) Biodigester effluent as fertilizer for water spinach established from seed or from cuttings. Livest Res Rural Dev 16(10)

Cann I, Kobayashi Y, Wakita M, Hoshino S (1994) Effects of three chemical treatments on in vitro fermentation of rice straw by mixed rumen microbes in the presence or absence of anaerobic rumen fungi. Reprod Nutr Dev 34(1):47–56

Carrere H, Antonopoulou G, Affes R, Passos F, Battimelli A, Lyberatos G, Ferrer I (2016) Review of feedstock pretreatment strategies for improved anaerobic digestion: from lab-scale research to full-scale application. Bioresour Technol 199:386–397

Carvalheiro F, Duarte LC, Gírio FM (2008) Hemicellulose biorefineries: a review on biomass pretreatments. J Sci Indust Res:849–864

Chandra R, Takeuchi H, Hasegawa T (2012) Methane production from lignocellulosic agricultural crop wastes: a review in context to second generation of biofuel production. Renew Sust Energ Rev 16(3):1462–1476

Chandra R, Takeuchi H, Hasegawa T, Vijay V (2015) Experimental evaluation of substrate's particle size of wheat and rice straw biomass on methane production yield. Agric Eng Int CIGR J 17(2):93–104

Chang VS, Holtzapple MT (2000) Fundamental factors affecting biomass enzymatic reactivity. Paper presented at the Twenty-first symposium on biotechnology for fuels and chemicals

Chen Y, Sharma-Shivappa RR, Chen C (2007) Ensiling agricultural residues for bioethanol production. Appl Biochem Biotechnol 143(1):80–92

Chen Y, Cheng JJ, Creamer KS (2008) Inhibition of anaerobic digestion process: a review. Bioresour Technol 99(10):4044–4064

Clark R, Speece R (1971) The pH tolerance of anaerobic digestion. Adv Water Pollut Res 1:1–13

Deublein D, Steinhauser A (2011) Biogas from waste and renewable resources : an introduction, 2nd, rev. and expanded ed. Wiley-VCH, Weinheim
Dinuccio E, Balsari P, Gioelli F, Menardo S (2010) Evaluation of the biogas productivity potential of some Italian agro-industrial biomasses. Bioresour Technol 101(10):3780–3783
Eastman JA, Ferguson JF (1981) Solubilization of particulate organic carbon during the acid phase of anaerobic digestion. J Water Pollut Cont Fed 53(3):352–366
Eder B, Krieg A, Schulz H (2006) Biogas-Praxis: Grundlagen, Planung, Anlagenbau, Beispiele, Wirtschaftlichkeit: Ökobuch-Verlag
Fang C, Boe K, Angelidaki I (2011) Anaerobic codigestion of desugared molasses with cow manure; focusing on sodium and potassium inhibition. Bioresour Technol 102(2):1005–1011
Gerardi MH (2003) The microbiology of anaerobic digesters. Wiley, Hoboken
Gharpuray M, Lee YH, Fan L (1983) Structural modification of lignocellulosics by pretreatments to enhance enzymatic hydrolysis. Biotechnol Bioeng 25(1):157–172
Ghosh A, Bhattacharyya B (1999) Biomethanation of white rotted and brown rotted rice straw. Bioprocess Eng 20(4):297–302
Gollakota K, Meher K (1988) Effect of particle size, temperature, loading rate, and stirring on biogas production from castor cake (oil expelled). Biol Wastes 24(4):243–249
González G, López-Santín J, Caminal G, Sola C (1986) Dilute acid hydrolysis of wheat straw hemicellulose at moderate temperature: a simplified kinetic model. Biotechnol Bioeng 28(2):288–293
Gu Y, Chen X, Liu Z, Zhou X, Zhang Y (2014) Effect of inoculum sources on the anaerobic digestion of rice straw. Bioresour Technol 158:149–155
Gunnerson CG, Stuckey DC, Greeley M, Skrinde R (1986) Anaerobic digestion: principles and practices for biogas systems. World Bank technical paper No. 49. The World Bank, Washington DC, United State
Gurung J B (1997) Review of literature on effects of slurry use on crop production. The Biogas Support Programme, Kathmandu, Nepal
Haruta S, Cui Z, Huang Z, Li M, Ishii M, Igarashi Y (2002) Construction of a stable microbial community with high cellulose-degradation ability. Appl Microbiol Biotechnol 59(4–5):529–534
Haug RT, Stuckey DC, Gossett JM, McCarty PL (1978) Effect of thermal pretreatment on digestibility and dewaterability of organic sludges. J Water Pollut Contr Feder 50(1):73–85
Hendriks AT, Zeeman G (2009) Pretreatments to enhance the digestibility of lignocellulosic biomass. Bioresour Technol 100(1):10–18
Hills DJ (1981) Anaerobic digestion of dairy manure and field crop residues. Agric Wastes 3:179–189
Jönsson O, Polman E, Jensen JK, Eklund R, Schyl H, Ivarsson S (2003) Sustainable gas enters the European gas distribution system. Raport Danish Gas Technology Center
Jördening H-J, Winter J (2006) Environmental biotechnology: concepts and applications. Wiley, New York
Kargi F, Dincer AR (1996) Effect of salt concentration on biological treatment of saline wastewater by fed-batch operation. Enzym Microb Technol 19(7):529–537
Kaur K, Sehgal GK, Sehgal H (1987) Efficacy of biogas slurry in carp, *Cyprinus carpio* var. *communis* (Linn.), culture-effects on survival and growth. Biol Wastes 22(2):139–146
Koster I, Lettinga G (1988) Anaerobic digestion at extreme ammonia concentrations. Biol Wastes 25(1):51–59
Kugelman IJ, McCarty PL (1965) Cation toxicity and stimulation in anaerobic waste treatment. J Water Poll Control Federation:97–116
Laanbroek H (1990) Bacterial cycling of minerals that affect plant growth in waterlogged soils: a review. Aquat Bot 38(1):109–125
Lee DH, Behera SK, Kim JW, Park HS (2009) Methane production potential of leachate generated from Korean food waste recycling facilities: a lab-scale study. Waste Manag 29:876–882
Lei Z, Chen J, Zhang Z, Sugiura N (2010) Methane production from rice straw with acclimated anaerobic sludge: effect of phosphate supplementation. Bioresour Technol 101(12):4343–4348

Li D, Liu S, Mi L, Li Z, Yuan Y, Yan Z, Liu X (2015) Effects of feedstock ratio and organic loading rate on the anaerobic mesophilic codigestion of rice straw and cow manure. Bioresour Technol 189:319–326

Lianhua L, Dong L, Yongming S, Longlong M, Zhenhong Y, Xiaoying K (2010) Effect of temperature and solid concentration on anaerobic digestion of rice straw in South China. Int J Hydrog Energy 35(13):7261–7266

Lim JS, Abdul Manan Z, Wan Alwi SR, Hashim H (2012) A review on utilisation of biomass from rice industry as a source of renewable energy. Renew Sust Energ Rev 16(5):3084–3094

Liu S, Wu S, Pang C, Li W, Dong R (2014) Microbial pretreatment of corn stovers by solid-state cultivation of *Phanerochaete chrysosporium* for biogas production. Appl Biochem Biotechnol 172(3):1365–1376

Makádi M, Tomócsik A, Orosz V (2012) Digestate: a new nutrient source: review. In: Biogas: IntechOpen

Malik R, Singh R, Tauro P (1987) Effect of inorganic nitrogen supplementation on biogas production. Biol Wastes 21(2):139–142

McCarty PL (1964) Anaerobic waste treatment fundamentals. Public Works 95(9):107–112

Menardo S, Airoldi G, Balsari P (2012) The effect of particle size and thermal pretreatment on the methane yield of four agricultural by-products. Bioresour Technol 104:708–714

Møller HB, Sommer SG, Ahring BK (2004) Methane productivity of manure, straw, and solid fractions of manure. Biomass Bioenergy 26(5):485–495

Momayez F, Karimi K, Horváth IS (2018) Enhancing ethanol and methane production from rice straw by pretreatment with liquid waste from biogas plant. Energy Convers Manag 178:290–298

Monnet F (2003) An introduction to anaerobic digestion of organic wastes. Remade Scotland

Moorhead K, Nordstedt R (1993) Batch anaerobic digestion of water hyacinth: effects of particle size, plant nitrogen content, and inoculum volume. Bioresour Technol 44(1):71–76

Mosier N, Wyman C, Dale B, Elander R, Lee Y, Holtzapple M, Ladisch M (2005) Features of promising technologies for pretreatment of lignocellulosic biomass. Bioresour Technol 96(6):673–686

Mshandete A, Björnsson L, Kivaisi AK, Rubindamayugi MS, Mattiasson B (2006) Effect of particle size on biogas yield from sisal fibre waste. Renew Energy 31(14):2385–2392

Mustafa AM, Poulsen TG, Sheng K (2016) Fungal pretreatment of rice straw with *Pleurotus ostreatus* and *Trichoderma reesei* to enhance methane production under solid-state anaerobic digestion. Appl Energy 180:661–671

Neyens E, Baeyens J, Creemers C (2003) Alkaline thermal sludge hydrolysis. J Hazard Mater 97(1–3):295–314

Ngan NVC (2012) Promotion of biogas plant application in the Mekong Delta of Vietnam. PhD dissertation. Technical University of Braunschweig

Ngan NVC, Viet LH, Cu ND, Phong NH (2011) Biogas production of pig manure with water hyacinth juice from batch anaerobic digestion. In: Environmental Change and Agricultural Sustainability in the Mekong Delta, pp 355–369

Nguyen VCN, Tran SN (2015) Greenhouse gas emmission from on-field straw burning in the Mekong Delta of Vietnam. In: Proceedings of 8th Asian Crop Science Association Conference, pp 43–50

Nguyen VCN, Phan NL, Nguyen TNL, Pham CM, Kieu TN (2015) Cobenefits from applying codigester's bio-slurry to farming activities in the Mekong Delta. Health Environ 1:30–44

Nguyen VH, Castalone A, Jamieson C, Gummert M (2016) Small-scale batch anaerobic digestion of rice straw. https://waset.org/pdf/books/?id=54846&pageNumber=1

Nianguo L (1984) Biogas in China. Trends Biotechnol 2(3):77–79

Nozhevnikova A, Kotsyurbenko O, Parshina S (1999) Anaerobic manure treatment under extreme temperature conditions. Water Sci Technol 40(1):215–221

Ogata Y, Ishigaki T, Nakagawa M, Yamada M (2016) Effect of increasing salinity on biogas production in waste landfills with leachate recirculation: a lab-scale model study. Biotechnol Reports 10:111–116

Parkin G F, Lynch NA, Kuo WC, Van Keuren EL, Bhattacharya SK (1990) Interaction between sulfate reducers and methanogens fed acetate and propionate. Res J Water Pollut Contr Feder:780–788
Pilli S, Bhunia P, Yan S, LeBlanc R, Tyagi R, Surampalli R (2011) Ultrasonic pretreatment of sludge: a review. Ultrason Sonochem 18(1):1–18
Polprasert C, Koottatep T (2007) Organic waste recycling. IWA Publishing, London
Rasi S, Veijanen A, Rintala J (2007) Trace compounds of biogas from different biogas production plants. Energy 32(8):1375–1380
Reinhart D, Townsend T (1998) Landfill bioreactor design and operation. CRC Press LLC, Boca Raton
Sharma SK, Mishra I, Sharma M, Saini J (1988) Effect of particle size on biogas generation from biomass residues. Biomass 17(4):251–263
Shen F, Li H, Wu X, Wang Y, Zhang Q (2018) Effect of organic loading rate on anaerobic codigestion of rice straw and pig manure with or without biological pretreatment. Bioresour Technol 250:155–162
Singh KJ, Sooch SS (2004) Comparative study of economics of different models of family size biogas plants for state of Punjab, India. Energy Convers Manag 45(9–10):1329–1341
Sophea K, Preston T (2001) Comparison of biodigester effluent and urea as fertilizer for water spinach vegetable. Livest Res Rural Dev 13(6):125–127
Sophin P, Preston T (2001) Effect of processing pig manure in a biodigester as fertilizer input for ponds growing fish in polyculture. Livest Res Rural Dev 13(6)
Spiegel RJ, Preston JL (2003) Technical assessment of fuel cell operation on anaerobic digester gas at the Yonkers, NY, wastewater treatment plant. Waste Manag 23(8):709–717
Stuckey DC, McCarty PL (1984) The effect of thermal pretreatment on the anaerobic biodegradability and toxicity of waste activated sludge. Water Res 18(11):1343–1353
Sun Y, Cheng J (2002) Hydrolysis of lignocellulosic materials for ethanol production: a review. Bioresour Technol 83(1):1–11
Sun C, Liu R, Cao W, Yin R, Mei Y, Zhang L (2015) Impacts of alkaline hydrogen peroxide pretreatment on chemical composition and biochemical methane potential of agricultural crop stalks. Energy Fuel 29(8):4966–4975
Taherzadeh M, Karimi K (2008) Pretreatment of lignocellulosic wastes to improve ethanol and biogas production: a review. Int J Mol Sci 9(9):1621–1651
Thanh PV (2010) VAC integrated system with entire energy chain in Vietnam. Paper presented at the FAO technical consultation conference
Tran SN, Vo TV, Nguyen HC, Nguyen VCN, Le HV, Ingvorsen K (2014) Enhancing biogas production by supplementing rice straw. J Sci Technol 52(3A):294–301
Tran SN, Huynh VT, Huynh CK, Nguyen VCN, Nguyen HC, Le HV, Kjeld I (2015) Evaluation the possibility of using rice straw and water hyacinth in semi continuous anaerobic fermentation: the application on farm-scale polyethylene biogas digesters. Can Tho University J Sci 36:27–35. (in Vietnamese)
Tran SN, Nguyen VCN, Nguyen HC, Le HV, Ingvorsen K (2017) Rice straw pretreatment methodology for biogas production. In: Rice straw in the Mekong Delta of Vietnam and its potential for biogas production. p 61–84 (in Vietnamese)
Tripathi A (1993) Biogas slurry: a boon for agriculture crops. Biogas slurry utilisation. CORT, New Delhi, pp 11–14
Wang Y, Zhang Y, Meng L, Wang J, Zhang W (2009) Hydrogen–methane production from swine manure: effect of pretreatment and VFAs accumulation on gas yield. Biomass Bioenergy 33(9):1131–1138
Ward AJ, Hobbs PJ, Holliman PJ, Jones DL (2008) Optimisation of the anaerobic digestion of agricultural resources. Bioresour Technol 99(17):7928–7940
Weemaes MP, Verstraete WH (1998) Evaluation of current wet sludge disintegration techniques. J Chem Technol Biotechnol: Intl Res Proc Environ Clean Technol 73(2):83–92

Wiese J, König R (2009) From a black-box to a glass-box system: the attempt towards a plant-wide automation concept for full-scale biogas plants. Water Sci Technol 60(2):321–327

Yadav M, Singh A, Balan V, Pareek N, Vivekanand V (2019) Biological treatment of lignocellulosic biomass by *Chaetomium globosporum*: process derivation and improved biogas production. Int J Biol Macromol 128:176–183

Yadvika S, Sreehrishnan TR, Kohli S, Rana V (2004) Enhancement of biogas production from solid substrates using different techiniques – a review. Bioresour Technol 95:1–10

Yan Z, Song Z, Li D, Yuan Y, Liu X, Zheng T (2015) The effect of initial substrate concentration, C/N ratio and temperature on solid-state anaerobic digestion for composting straw. Bioresour Technol 177:266–273

Ye J, Li D, Sun Y, Wang G, Yuan Z, Zhen F, Wang Y (2013) Improved biogas production from rice straw by codigestion with kitchen waste and pig manure. Waste Manag 33(12):2653–2658

Zhang R, Zhang Z (1999) Biogasification of rice straw with an anaerobic-phased solids digester system. Bioresour Technol 68(3):235–245

Zhao J, Ge X, Vasco-Correa J, Li Y (2014a) Fungal pretreatment of unsterilized yard trimmings for enhanced methane production by solid-state anaerobic digestion. Bioresour Technol 158:248–252

Zhao J, Zheng Y, Li Y (2014b) Fungal pretreatment of yard trimmings for enhancement of methane yield from solid-state anaerobic digestion. Bioresour Technol 156:176–181

Ziemiński K, Frąc M (2012) Methane fermentation process as anaerobic digestion of biomass: transformations, stages, and microorganisms. Afr J Biotechnol 11(18):4127–4139

Zinder SH (1993) Physiological ecology of methanogens. In: Methanogenesis, pp 128–206. Springer, Berlin

Chapter 6
Rice-Straw Mushroom Production

Le Vinh Thuc, Rizal G. Corales, Julius T. Sajor, Ngo Thi Thanh Truc, Phan Hieu Hien, Remelyn E. Ramos, Elmer Bautista, Caesar Joventino M. Tado, Virgina Ompad, Dang Thanh Son, and Nguyen Van Hung

Abstract The rice-straw mushroom *(Volvariella volvacea)* has a distinct flavor, pleasant taste, and rich protein content. It has low production costs and a cropping duration of approximately 45 days—making it an effective means for poverty alleviation for those farmers who grow it. Farmers in Vietnam, the Philippines, and Cambodia grow it. Rice straw is one of the most common substrates used for growing this mushroom. The mushroom can grow well in both outdoor and indoor conditions; however, outdoor cultivation has risks of exposure to rain, wind, and/or high temperatures, all which reduce yield. The yield of indoor mushroom production is higher and more stable, as such, indoor growing is preferred. In addition to cultivation, this chapter also covers straw mushroom characteristics, cultivation principles and techniques, and rice straw substrate preparation.

Keywords Rice-straw mushroom · Indoor cultivation · Outdoor cultivation

L. V. Thuc (✉) · N. T. T. Truc
Can Tho University, Can Tho, Vietnam
e-mail: lvthuc@ctu.edu.vn; ntttruc@ctu.edu.vn

R. G. Corales · J. T. Sajor · R. E. Ramos · E. Bautista · C. J. M. Tado · V. Ompad
Philippine Rice Research Institute (PhilRice), Muñoz, Philippines
e-mail: rg.corales@philrice.gov.ph; jt.sajor@philrice.gov.ph; re.ramos@philrice.gov.ph; eg.bautista@philrice.gov.ph; cjm.tado@philrice.gov.ph

P. H. Hien
Nong Lam University, Ho Chi Minh City, Vietnam

D. T. Son
University of Technical Education, Vinh Long, Vietnam

N. V. Hung
International Rice Research Institute (IRRI), Los Baños, Laguna, Philippines
e-mail: hung.nguyen@irri.org

M. Gummert et al. (eds.), *Sustainable Rice Straw Management*,
https://doi.org/10.1007/978-3-030-32373-8_6

6.1 Overview of Rice-Straw Mushroom (RSM)

Mushroom is considered an important food to address food and nutrition security and human health (Ishara et al. 2018; Cuesta and Castro-Rios 2017; Feeney et al. 2014a) and climate change adaptation issues (Gellerman 2018; Langston 2014). *Volvariella volvacea* (Fig. 6.1), also known as the straw mushroom or rice-straw mushroom (RSM), is one species of edible mushroom cultivated throughout East and Southeast Asia (Sudha et al. 2008).

RSM production adds value to rice production and increases the income of the poor farmers in developing countries (Imtiaj and Rahman 2008; Shakil et al. 2014; Zhang et al. 2014).

Among more than 38,000 known mushroom species, such as *Agaricus bisporus*, *Lentinus edodes*, *Flammulna velutipis, Auricularia polytricha*, etc., RSM is one of the most common mushrooms cultivated (Walde et al. 2006) and ranks third among important mushrooms due to its delicious taste (Ramkumar et al. 2012; Thiribhuvanamala et al. 2012), as well as its short growing time compared to other species (Rajapakse 2011). In terms of production, RSM ranks sixth among edible mushrooms, accounting for about 5–6% of world production (Ahlawat et al. 2011).

RSM is known as a healthy food (Belewu and Belewu 2005; Feeney et al. 2014b; USITC 2010). It has high protein, potassium, and phosphorus contents (Ahlawat and Tewari 2007) while being salt-free and low in alkalinity, fat, and cholesterol. Mushroom also contains selenium (Solovyev et al. 2018) and niacin (Ahlawat and Tewari 2007; Eguchi et al. 2015), which are two essential compounds in the immune system and the thyroid that have a role in cancer prevention (Hobbs 1995). Its fiber content is important for physiological functions in the gastrointestinal tract (Manzi et al. 2001). In addition, RSM has significant antimicrobial activity (Chandra and Chaubey 2017). It also provides good sources of polypeptide, terpene, and steroid (Shwetha and Sudha 2012) and phenolic compounds, such as flavonoids, phenolic

Fig. 6.1 *Volvariella volvacea* (Bull.; Fr.) Singer is an edible mushroom also known as the straw mushroom and, for our purposes, the rice-straw mushroom (RSM)

Table 6.1 Chemical composition of RSM

Content	Unit	Value
Moisture content	% in wet basis	88–91.1
Crude protein	% of dry weight	30–43
Crude fat	% of dry weight	1–6
Fiber	% of dry weight	4–10
Ash	% of dry weight	5–13
Carbohydrate	% of dry weight	12–49.3
Energy value	Kcal/kg of dry weight	2760

Adapted from Chang and Quimio (1989), Eguchi et al. (2015)

acid, and tannins that contribute to its high antioxidant properties (Hung and Nhi 2012). Other sources of antioxidants in RSM are catalase, superoxide dismutase, glutathione peroxidase, peroxidase, glutathione-S-transferase, and glutathione reductase (Ramkumar et al. 2012). Table 6.1 summarizes the chemical composition of RSM.

Due to its many benefits and advantages, mushroom production and consumption have significantly increased in many countries (Vizhanyo and Jozsef 2000; Bernaś et al. 2006). The top mushroom producers are China, USA, and The Netherlands, contributing 47%, 11%, and 4%, respectively of the world's total mushroom production.

6.2 Physical Characteristics of RSM

RSM is best adapted in tropical and subtropical regions (Bao et al. 2013) and grows at relatively high temperatures (Obodai and Odamtten 2012). Its total crop cycle, under favorable growing conditions, is within 4–5 weeks (Biswas 2014). It belongs to the Fungi kingdom, Plutaceae family, Agaricales order, Agaricomycetes class, and Basidiomycota division (Chang 1969, 1974; Rajapakse 2011). RSM has an umbrella-shaped cap (pileus) ranging from dark grey to brown and a diameter of 8–10 cm. When young, its cap has an egg-like shape and, as it matures, it becomes cone-like and nearly flat. The stalk (stipe) ranges in color from silky white to brown, which develops to a brownish gray sack-like cup (volva) (Chang and Miles 2004). The mycelia, the vegetative parts, comprise of threads and cord-like strands branching out through the substrate.

When the mycelia come together, the mushroom begins its first stage of development called the ***pinhead stage.*** It is characterized by tiny clusters in white circular structures of interwoven thread-like hyphae. This is followed by the ***button stage*** in which buttons encircling the egg-shape structures are covered by a layer of tissue or a universal veil (volva). The stalk (stipe), cap (pileus), and gills (lamellae) are seen inside the button when it is cut lengthwise. Commercially, the button stage is pre-

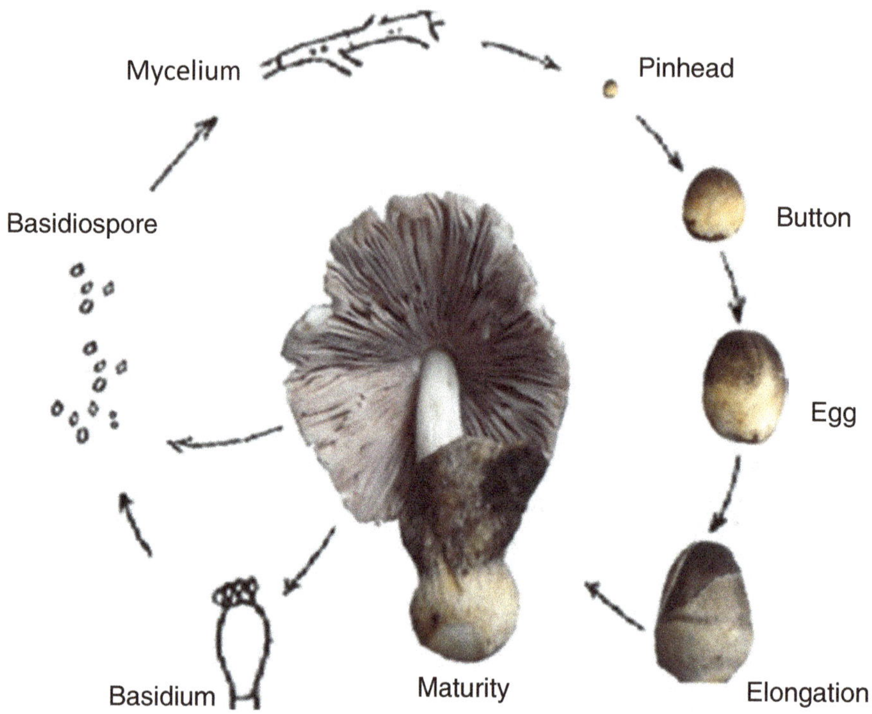

Fig. 6.2 RSM life cycle. (Adapted from Le-Duy-Thang 2006; Bao et al. 2013)

ferred for harvesting because of the mushroom's high-protein content at this point (about 25%), best palatability, and longer shelf life. The ***elongation stage*** occurs after the universal veil (volva) ruptures, exposing the stalk and the cap. The last stage, ***maturity,*** is characterized by the fully expanded cap exposing the brownish-pink gills of its lower surface. At this stage, the spawns (basidiospores) begin to discharge. Figure 6.2 shows the mushroom lifecycle starting from generation of the spawns and ending with the formation of the ear.

6.3 Environmental and Nutritional Requirements

RSM is considered as one of the easiest mushrooms to cultivate because of its short production duration (Zikriyani et al. 2018) and advantages of having less fat. As mentioned, this species grows in warm weather, typically in the tropics and subtropics. The optimal temperature is from 30 to 35 °C for the RSM's mycelial growth and from 28 to 30 °C for its fruiting body production (Le-Duy-Thang 2006). The suitable temperature for growing mushrooms is between 25 and 40 °C with the optimum being 35 °C (Fasidi 1996). Relative humidity in the range of from 70% to 90% is best for RSM growth (Biswas and Layak 2014). The optimal pH is 6.5; anything

Table 6.2 Environmental requirements for RSM growth

	Mycelium		Fruiting	
Parameter	Range	Optimal	Range	Optimal
Temperature, °C	15–42	35 ± 2	25–30	28 ± 2
Relative humidity, %	50–70	60 ± 5	80–100	90 ± 5
pH	6–7	6.5	6–7	6.5

Adapted from Fasidi (1996), Chang and Miles (2004), Akinyele and Adetuyi (2005), Le-Duy-Thang (2006), Biswas and Layak (2014)

higher hampers mycelia growth (Akinyele and Adetuyi 2005). This species grows well on a number of cellulosic substrates, such as rice straw, wheat straw, sugarcane bagasse, banana leaves, water hyacinth, etc. RSM production can be intensified with the development of cutting-edge technologies. It can be grown outdoors or indoors. Growing practices are described Sect. 6.4. Table 6.2 summarizes the main parameters that enhance RSM growth.

Traditionally, RSM is mostly cultivated outdoors because of the low investment cost. However, outdoor cultivation has low and unstable productivity due to exposure to changing weather conditions (Reyes 2000). Although controlled indoor mushroom cultivation requires more investment, it usually results in higher and more stable yields (Chang 1996). In addition, through environmental control, RSM can be intensively cultivated, growing from six to eight crops annually. Palitha (2011) reported that the yield of indoor RSM cultivation can be 2.7 times higher than that of outdoor practice with the same application of feedstock.

Biological efficiency (BE) is an important parameter used in the mushroom industry to evaluate the effectiveness of a mushroom strain on different substrates (Chang et al. 1981; Biswas and Layak 2014; Girmay et al. 2016). It is calculated as follows:

$$\mathrm{BE} = \frac{FWm}{DWs} * 100\%$$

where:

BE is the biological efficiency
FWm is the total fresh weight (g) of mushroom yield across all flushes, and
DWs is the substrate dry weight (g)

As already mentioned, RSM can be cultivated on several lignocellulose materials; however, RSM productivity is attributed to substrates of the best quality (Ahlawat et al. 2011). Table 6.3 shows the biological efficiency of RSM production on different substrates.

Table 6.3 Biological efficiency of RSM on different substrates

Substrates	Biological efficiency (%)	References
Rice straw	10.2–15.0	Chang et al. (1981), Biswas (2014), Chang and Miles (2004)
Cotton waste	17.7–40.0	Chang et al. (1981), Zikriyani et al. (2018)
Banana leaves	8.6–15.2	Belewu and Belewu (2005), Zikriyani et al. (2018)
Sugarcane trash	13.2	Thiribhuvanamala et al. (2012)
Wheat	12.1	Tripathy (2010)
Water hyacinth	8.7	Thiribhuvanamala et al. (2012)

6.4 Current Practices for Growing Mushroom

6.4.1 Outdoor RSM Cultivation

The steps for outdoor RSM production (Fig. 6.3), as it is done in Vietnam, are shown in Fig. 6.4.

6.4.1.1 Rice Straw for Mushroom Growing and Preparation of the Growing Location

Rice straw intended for growing RSM should be dry, clean, without mold contamination, and should not have been exposed to rain or should not have started rotting in the field. Rice straw contaminated with molds may have mycelia or spawns with a white color. To minimize contamination and for best quality, the straw should be collected right after harvest. The location for growing RSM should be cleaned and treated with 300–500 kg ha^{-1} (3–5 kg 100 m^{-2}) of lime ($CaCO_3$) 3 days before incubation.

6.4.1.2 Growing Preparation and Maintenance of Planting Spawn

The most commonly used spawn substrate is a mixture of tobacco midrib and sawdust. The tobacco midribs are first soaked in clean water overnight. After soaking, they are washed at least three times, drained, and then chopped into lengths of from 2 to 4 cm. The chopped midribs are boiled for 30 minutes and then drained until the moisture content reaches around 65%. Next, the midribs are mixed with the sawdust. About 350 g of mixed spawn substrate is placed inside a 6- × 12-in. polypropylene (PP) bag. For easier handling, a plastic ring may be placed as a “bottle neck” on the PP bag. This can be done by pulling out the PP bag end through the polyvinylchloride (PVC) ring then folding the pulled-out part outward to make an open-

Fig. 6.3 Growing RSM outdoors

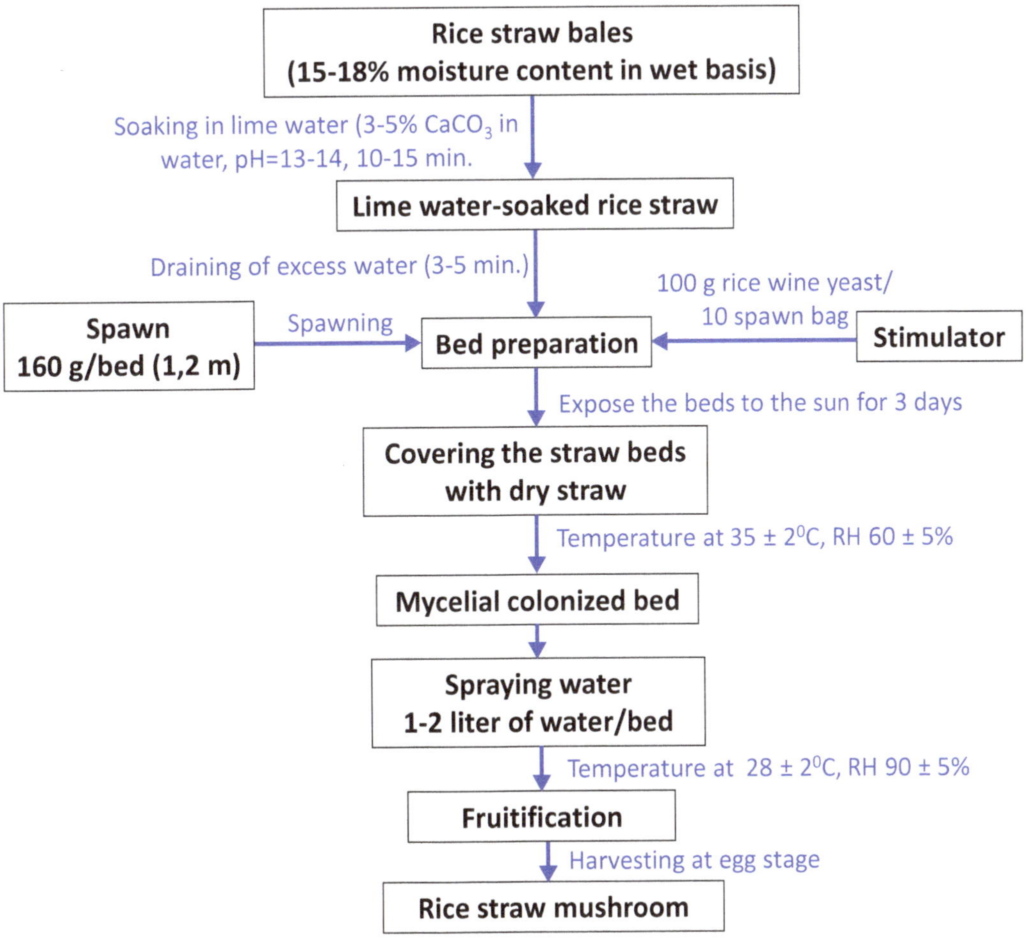

Fig. 6.4 Process of RSM production from rice straw preparation to mushroom harvesting. (Adapted from Thuc et al. 2019)

ing. Next, the folded part is secured by tying with a rubber band. The PVC neck opening is plugged with a rolled cotton waste then covered with paper secured with rubber band.

The PP bags containing the spawn substrates are sterilized using an autoclave at 15 psi pressure and 121 °C for 30 min. The sterilized PP bags with the spawn substrates are then transferred to the inoculation room and allowed to cool down.

The sterilized bags are kept inside the laminar flow under a UV tube or inoculation chamber for 20–30 min. Inoculation is done by removing the cotton plug of each bag, then placing a 1-sq mm pure culture mycelial block on top of the spawn substrate using a sterilized inoculation needle, then replacing the cotton plug. The process is repeated until all the bags with substrates have been inoculated.

The inoculated bags are kept in the incubation room at 32 °C temperature for 2 weeks, or until mycelial growth reaches the bottom of each bag. The bags should always be checked for contamination during the incubation period. The shelf life of the spawn is about 4 weeks at room temperature. It can also be refrigerated at 4 °C to prolong storage. The refrigerated spawn should be primed at room temperature before using in order to activate spawn growth.

6.4.1.3 Preparation of Growing Beds and Spawning

Rice straw or stubble can be used as bedding materials or substrates. These materials collected from the field must be sun-dried. If bundled substrates are used as bedding, the straw should be cut into 30-cm long strips to make bundles 10 cm in diameter. The beds can be created manually (Fig. 6.5a) or using a

Fig. 6.5a Manual bedding

Fig. 6.5b Bedding using a wooden frame

wooden frame (Fig. 6.5b). The wooden frame size is 0.3–0.4 m in width, 0.35–0.4 m in height, and 1.5 m in length. Straw should be placed into the frame and compacted so that the first layer is 10 cm thick; then the spawn is added to the straw surface. A second layer using similar steps should be done. The two layers of straw are compressed, and then the frame is removed to have the beds on the ground for growing RSM.

The bedding materials are soaked in clean water for 12 h to make them soft and pliable. The soaked substrates are rinsed with clean water to remove the slime, fermenting odor, and to reduce acidity. Soaking is a prelude to composting.

In composting, the soaked substrates are piled up then covered with plastic sheets. The composting period is 14 days and the pile should be turned on the 7th day to ensure even composting. In some cases, 1% molasses and 5% complete fertilizer (14–14-14 NPK) are mixed into the substrate during composting. Agricultural lime (1%) is also added when the compost pile is turned. Through composting, the substrates are converted into a rich medium suitable for mushroom growth.

The moisture content of the substrate during bed preparation must be close to 65%. Growing beds are established by piling the bundled substrates into layers. The spawns are sprinkled thinly over the bundles in each layer. It can also be placed in thumb-size bands 7 cm from the edge of the bed at a distance of 10 cm between bands. Sometimes the spawn is covered with newspaper to protect the spawn from drying and to enhance better mycelial growth. If the substrate were not applied with molasses and fertilizer during composting, a nutrient solution, containing 10 g of urea and 30 g of sugar mixed in 4 L of water, is sprinkled over each substrate layer. The process should be repeated until all layers have been treated. Ideally, the bed should have three layers and should be from 2.5 to 3 m long.

The growing bed is covered with a polyethylene plastic sheet to maintain the desired temperature and relative humidity appropriate for mycelial growth. The optimum temperature for incubation ranges from 30 to 35 °C with a relative humidity ranging from 75 to 85%. The incubation period takes from 10 to 14 days. Mushroom primordia or pinheads usually appear on the side and surface of the growing beds 5 days after spawning. Once pinheads are observed, the plastic sheet cover should be lifted for a while to introduce fresh air. The temperature should be maintained at 30 to 32 °C to synchronize fruiting body formation during the fruiting stage. The surroundings of the beds should be watered to help maintain the desired temperature.

6.4.1.4 Mushroom Growing Care

In the first 3 days after adding spawn to the straw beds, the beds need to be exposed to the sun to increase the temperature inside, which stimulates mycelial growth. Then, the beds are covered with a net and dry straw. Some nutritional supplements or stimulants such as Bioted, HQ, or HVP 301 can be sprayed onto the beds to enhance better mushroom growth. The beds can be watered and covered with rice straw to maintain the temperature and humidity as well as to maximize the yield and quality of RSM production, as indicated in Table 6.2.

6.4.1.5 Harvesting and Processing

The first fruiting flush occurs about 14 days after incubation and continues for about 5 days. After the fruiting flush, water is sprinkled over the bed and covered again with the plastic sheet to build up the temperature. Within 7–14 days, the next fruiting flush will appear. The succeeding fruiting flushes often consist of larger, but fewer fruiting bodies than the first flush. Hand picking is the common method of harvesting and sorting the mushrooms. This guarantees less damage and better quality. The mushrooms are picked from the growing beds with a rotating motion. The harvest is sorted based on quality and size. To enhance higher protein content, better palatability, and longer shelf life, the preferred times for harvesting are during the button to egg-shaped stages.

6.4.2 Indoor RSM Growing

Indoor mushroom growing requires the same preparation and treatment steps as in the outdoors. However, the environmental criteria, such as heap temperature (>70 °C) to sterilize straw, moisture content (60–65%), etc., have to be strictly controlled. Indoor RSM growing uses shelves with two types of bedding, spread (Fig. 6.6a) and compacted (Fig. 6.6b). The ratio of spawn used is about 200 g m^{-2}.

Fig. 6.6a Spread bedding

Fig. 6.6b Compacted bedding

It is necessary to cover the substrate beds to secure the moisture content for 2–3 days. Water may be sprinkled upon seeing the fungus grow on most of the beds. Organic fertilizer, such as chicken manure or cow dung, is added to the substrate at a rate of about 0.5–1.5 kg m^{-2} to increase the nutrient uptake by the mushrooms. All materials have to be sterilized before adding them to the substrate.

6.4.3 *Case Study of Cost-Benefits for Growing Indoor and Outdoor Mushroom*

We conducted assessments for indoor and outdoor mushroom growing in the Mekong River Delta (MRD) of Vietnam in 2018 that resulted in the cost-benefit comparison shown in Table 6.4. For the outdoor practice, total input cost was about 1.28 $US kg^{-1} of mushroom produced and 1.23 $US m^{-2} of land used. It comprises the main component costs of rice straw (40%), labor (23%), chemical inputs (11%), and the rest for land use, depreciation of net and pump, and watering. On the other hand, for the indoor practice, the total input cost was 1.37 $US kg^{-1} of mushroom produced and 10.79 $US m^{-2} of.

land used. The indoor practice cost breakdown was depreciation of growing house and facilities, 44%; rice straw, 31%; labor, 7%; and the rest for use, depreciation of net, pump, and growing house (for indoor scenario), and watering. Net profit accounted for 1 kg of mushroom produced was the same for both indoor and outdoor practices at 0.5–0.6 $US kg^{-1}. Whereas, accounting for a square meter of land

Table 6.4 Comparing cost-benefits between outdoor and indoor RSM growing practices in MRD

Parameters	Outdoor $US kg^{-1} of mushroom	Outdoor $US m^{-2} of land used	Indoor $US kg^{-1} of mushroom	Indoor $US m^{-2} of land used
Inputs				
Land used (rental)	0.15	0.04	0.16	0.35
Rice straw	0.51	0.38	0.54	3.33
Net, pump, depreciation of growing house (indoor only)	0.03	0.54	0.03	4.76
Lime, fertilizer and pesticide	0.12	0.07	0.13	0.60
Spawns	0.14	0.10	0.15	0.83
Watering (power consumption)	0.03	0.02	0.03	0.21
Labor	0.30	0.08	0.32	0.71
Total inputs	1.28	1.23	1.37	10.79
Outputs				
Mushroom	1.67	1.67	1.78	14.58
Spent rice straw	0.15	0.10	0.16	0.83
Total outputs	2.35	2.29	2.51	20.04
Net profit	0.5	0.6	0.5	4.6

used, net profit of the indoor practice was 4.6 \$US m^{-2} about 9 times higher than that of the outdoor practice. However, RSM is commonly cultivated in rural areas, near the rice fields to reduce the cost of transporting the rice straw. So, outdoor mushroom growing is still widely done in Vietnam.

6.5 Pest and Disease Problems

RSM is very sensitive to the environment including temperature, sunlight, water, oxygen (O_2), and carbon dioxide (CO_2). Sudden changes in temperature may hamper or even stop mushroom growth. Sunlight is needed from the sphere to the egg stages. With a lack of sunlight, vitamin E will be significantly reduced, vitamin D will not be available, and melanin pigment (black pigment) will not form in RSM.

Green mold (*Verticillium fungicola*), orange mold (*Neurospora* spp.), plaster mold (*Scopulariopsis fimicola*), acne mushroom (*Selerotium rolfsii*), etc. are the typical diseases that affect RSM. These diseases can be prevented or treated by using lime water with a 0.5–1% concentration and applied by watering on the affected area. Gypsum disease can be treated with potassium permanganate ($KMnO_4$) or acetic acid (40%). If the disease is severe, it can be treated by fungicides, such as Benomyl 0.1%, 7% Zineb, or Validacin (for acne).

6.6 Preservation and Consumption of RSM

RSM can be used and processed into many different products but it is easily damaged during harvesting and primary processing. The selection of appropriate technology for product storage and processing on a scale that is compatible with production conditions will promote the cultivation of mushrooms and help stabilize consumption.

RSM spoils very quickly and can be stored at most for 3 days at temperatures between 10 and 15 °C or in controlled atmosphere packaging (Jamjumroon et al. 2012) it loses moisture in 4 days, resulting in a 40–50% loss of mushroom weight when stored under normal ambient temperature. Thus, other methods are used for longer storage, one of which is dried RSM. However, sun drying often changes the color and taste of the product. Furthermore, RSM exposed to the sun outdoors is susceptible to microbial contamination. The drying process takes 24 h at 30 °C. The drying temperature can start at 40 °C and then gradually increase over 8 h to 45 °C. Raw materials of dried mushrooms can be left or cut in half. If cut in half, they must be pretreated before drying. Blanching for 3–4 min in hot water or 4–5 min in hot steam helps mushrooms keep their color better during storage. When RSM is dried at 60 °C for 7 h, the moisture content may reach 5%. Dried mushrooms can be stored or pulverized for use in spices. Other methods recommended for RSM preservation include air-conditioning packaging with storage media

(Lopez-Briones et al. 1992), drying (Izli and Isik 2014), freezing (Murr and Morris 1975), soaking in saline or acid solution (Cliffe-Byrnes and O'Beirne 2008), and canning (Vivar-Quintana et al. 1999).

Storage time can be extended for 3–6 months by soaking the mushrooms in acidic or saline solutions, which help extend shelf life and maintain their color. The mushrooms are washed in plain water before dipping into the saline solution. The mushrooms are then put in the containers and covered with the saline solution.

Mushroom preservation through industrial canning technology is used in many countries around the world. The process of producing canned RSM includes preliminary processing, blanching, stacking, sterilization, cooling, labeling, and packaging. In order to produce canned mushrooms of good quality, it is necessary to process harvested mushrooms as soon as possible. In case of unavoidable delay, mushrooms should be stored at 4–5 °C until processed.

However, all the other preservation methods result in inferior mushroom eating quality compared to that of fresh mushroom, in terms of the original flavor, color, hardness, and so on. Extending the shelf life of fresh mushroom beyond 3 days is most important, as illustrated in the case of the Mekong Delta in Vietnam. In the local market, mushrooms are consumed as a fresh vegetable with the price normally fluctuating from 2 to 4 US\$ kg^{-1} at the first and 15th day of the lunar month. A small portion of salted or dried RSM is also exported at 2 US\$ kg^{-1}, but is not as much appreciated as fresh mushrooms. For estimating consumer trends, we can look at the American market. In 2012, the share of fresh mushrooms was 87% in quantity and 93% in value; the remaining minor portion is processed mushroom, with a farm gate price of only one half compared to that of fresh mushroom (Phan-Hieu-Hien 2017).

The price of fresh RSM at US supermarkets in 2013 was about 10 \$US kg^{-1}, while that of salted mushroom was only 5 \$US kg^{-1} (personal communication with Mr. Le Duy Thang, mushroom expert). From farms in Vietnam to US supermarkets, fresh RSM needs a minimum of 8 days to "travel", including 2–3 days through customs and 2–3 days at supermarkets before reaching consumers. The 8-day shelf life of fresh mushroom is the greatest constraint to boost mushroom production, or indirectly to increase the use of rice straw. Luckily after decades of deadlock, some research results are promising (Dhalsamant et al. 2018). Factors to help ensure a successful 8-day storage cycle include: (1) a suitable temperature, say 12 °C; (2) a controlled-atmosphere packaging, which is balanced between oxygen and carbon dioxide content; and (3) a chemical pretreatment, such as $CaCl_2$. More in-depth research is needed in parallel with pilot testing for economic performance.

6.7 Summary and Recommendations

Producing RSM is a sustainable option for adding value to rice production and reducing environmental harm through avoiding the burning of rice straw in the field. Growing outdoor RSM is a traditional practice with low investment costs but generates low yield and incurs high risk because it is strongly affected by changes in the weather. On the other hand, growing indoor RSM has higher investment costs but greater productivity and lower risks due to its well controlled environment.

One of the major bottlenecks for developing RSM is its market. Even though fresh RSM has high value, it cannot be stored for more than 3 days because it is highly perishable. Using technology to improve preservation to lengthen the storage time is a key to increasing the market and price and improving RSM's value chain.

References

Ahlawat OP, Tewari RP (2007) Cultivation technology of paddy straw mushroom (*Volvariella volvacea*). Technical Bulletin. Yugantar Prakashan Pvt. Ltd. WH-23, Mayapuri Indl. Area, New Delhi. p 64

Ahlawat OP, Singh R, Kumar S (2011) Evaluation of *Volvariella volvacea* strains for yield and diseases/insect-pests resistance using composted substrate of paddy straw and cotton mill wastes. Indian J Microbiol 51(2):200–205

Akinyele BJ, Adetuyi FC (2005) Effect of agrowastes, pH and temperature variation on the growth of *Volvariella volvacea*. Afr J Biotechnol 4(12):1390–1395

Bao D, Gong M, Zheng H, Chen M, Zhang L, Wang H, Jiang J, Wu L, Zhu Y, Zhu G, Zhou Y, Li C, Wang S, Zhao Y, Zhao G, Tan Q (2013) Sequencing and comparative analysis of the straw mushroom (*Volvariella volvacea*) genome. PLoS One 8(3):e58294. https://doi.org/10.1371/journal.pone.0058294

Belewu MA, Belewu KY (2005) Cultivation of mushroom (*Volvariella volvacea*) on banana leaves. Afr J Biotechnol 4(12):1401–1403

Bernaś E, Jaworska G, Kmiecik W (2006) Storage and processing of edible mushrooms. Acta Sci Pol Technol Aliment 5:5–23

Biswas MK (2014) Cultivation of paddy straw mushrooms (*Volvariella volvacea*) in the lateritic zone of West Bengal-a healthy food for rural people. Intl J Econ Plants 1(1):043–047

Biswas MK, Layak M (2014) Techniques for increasing the biological efficiency of paddy straw mushroom (*Volvariella Volvacea*) in eastern India. Food Sci Technol 2(4):52–57

Chandra O, Chaubey K (2017) *Volvariella volvacea*: a paddy straw mushroom having some therapeutic and health prospective importance. World J Pharm Pharm Sci 6(9):1291–1300

Chang ST (1969) A cytological study of spore germination of *Volvariella volvacea*. Bot Mag 82:102–109

Chang ST (1974) Production of straw mushroom (*Volvariella volvacea*) from cotton wastes. Mushroom J 21:348–354

Chang ST (1996) Mushroom research and development - equality and mutual benefit. Proceedings of the 2nd international conference on mushroom biology and mushroom products. Pennsylvania State University, Pennsylvania, pp 1–10

Chang ST, Miles PG (2004) Mushrooms: cultivation, nutritional value, medicinal effect, and environmental impact. CRC Press, Boca Raton. 480 p

Chang ST, Quimio TH (1989) Tropical mushrooms: biological nature and cultivation methods. The Chinese University of Hong Kong, Hong Kong

Chang ST, Lau OW, Cho KY (1981) The evaluation and nutritive value of Pleurotus sajor-caju. European J Appl Microbiol Biotechnol 12:58–62

Cliffe-Byrnes V, O'beirne D (2008) Effects of washing treatment on microbial and sensory quality of modified atmosphere (MA) packaged fresh sliced mushroom (*Agaricus bisporus*). Postharvest Biol Technol 48:283–294

Cuesta MC, Castro-Rios K (2017) Mushroom as strategy to reduce food insecurity in Cambodia. Nutr Food Sci 47(6):817–828

Dhalsamant K, Dash SK, Bal LM, Panda MK (2018) Evaluation of modified atmosphere packaged chemically treated mushroom (*Volvariella volvacea*) for shelf life quality. Agric Eng Today 42(1):66–72

Eguchi F, Kalaw SP, Dulay RMR, Miyasawa N, Yoshimoto H, Seyama T, Reyes RG (2015) Nutrient composition and functional activity of different stages in the fruiting body development of Philippine paddy straw mushroom, *Volvariella volvacea* (Bull.:Fr.) Sing. Adv Environ Biol 9(22):54–65

Fasidi IO (1996) Studies on *Volvariella esculenta* (mass) singer: cultivation on agricultural wastes and proximate composition of stored mushrooms. Food Chem 55(2):161–163

Feeney MJ, Dwyer J, Hasler-Lewis CM, Milner JA, Noakes M, Rowe, S, Wach M, Beelman RB, Caldwell J, Cantorna MT, Castlebury LA, Chang ST, Cheskin LJ, Clemens R, Drescher G, Fulgoni III VL, Haytowitz DB, Hubbard VS, Law D, Miller AM, Minor B, Percival SS, Riscuta G, Schneeman B, Thornsbury S, Toner CD, Woteki CE, Wu D (2014a) Mushrooms and health summit proceedings. Am Soc Nutr https://doi.org/10.3945/jn.114.190728

Feeney MJ, Miller AM, Roupas P (2014b) Mushrooms – biologically distinct and nutritionally unique exploring a "third food kingdom". Nutr Today 49(6):301–307

Gellerman B (2018) Lowly in stature, fungi play a big role in regulating climate. https://www.wbur.org/news/2018/09/18/mushrooms-fungi-climate

Girmay Z, Gorems W, Birhanu G, Zepdie S (2016) Growth and yield performance of *Pleurotus ostreatus* (Jacq. Fr.) Kumm (oyster mushroom) on different substrates. AMB Express 6:87.. Published online 2016 Oct. 4. https://doi.org/10.1186/s13568-016-0265-1

Hobbs H (1995) Medicinal mushrooms. Amer IBot 87:821–827. https://www.researchgate.net/publication/324080400_Edible_mushrooms_new_food_fortification_approach_toward-food-security

Hung PV, Nhi NNY (2012) Nutritional composition and antioxidant capacity of several edible mushrooms grown in the southern Vietnam. Int Food Res J 19(2):611–615

Imtiaj A, Rahman SA (2008) Economic viability of mushrooms cultivation to poverty reduction in Bangladesh. Trop Subtrop Agroecosyst 8:93–99

Ishara J, Kenji GM, Sila DN (2018) Edible mushrooms: new food fortification approach towards food security. LAP Lambert Academic Publishing, Congo. http://www.tropentag.de/2018/abstracts/links/Jackson_yTIhZlpj.pdf

Izli N, Isik E (2014) Effect of different drying methods on drying characteristics, colour and microstructure properties of mushroom. J Food Nutr Res 53(2):105–116

Jamjumroon S, Wongs-Aree C, McGlasson WB, Srilaong V, Chalermklin P, Kanlayanarat S (2012) Extending the shelf-life of straw mushroom with high carbon dioxide treatment. J Food Agric Environ 10(1):78–84

Langston A (2014) Can mushroom fight climate change? Greener Ideal https://greenideal.com/news/environment/0225-can-mushrooms-fight-climate-change/

Lopez-Briones G, Varoquaux P, Chambroy Y, Bouquant J, Bureau G, Pascat B (1992) Storage of common mushroom under controlled atmospheres. Int J Food Sci Technol 27:493–505

Manzi P, Aguzzi A, Pizzoferrato L (2001) Nutritional value of mushrooms widely consumed in Italy. Food Chem 73(3):321–325

Murr DP, Morris LL (1975) Effect of storage temperature on postharvest changes in mushroom. J Am Soc Hortic Sci 100(1):16–19

Obodai M, Odamtten GT (2012) Mycobiota and some physical and organic composition of agricultural wastes used in the cultivation of the mushroom *Volvariella volvacea*. J Basic Appl Mycol 3:21–26

Palitha R (2011) New cultivation technology for paddy straw mushroom (Volvariella volvacea). In: Proceedings of the 7th International Conference on Mushroom Biology and Mushroom Products (ICMBMP7)

Phan-Hieu-Hien (2017) Utilization of rice straw in the world and in Vietnam. J Agric Sci Technol 6:17–31

Rajapakse P (2011) New cultivation technology for paddy straw mushroom (*Volvariella volvacea*). In Proceedings of the 7th International Conference on Mushroom Biology and Mushroom Products (ICMBMP7). 446–451

Ramkumar L, Ramanathan T, Johnprabagaran J (2012) Evaluation of nutrients, trace metals and antioxidant activity in *Volvariella volvacea* (Bull. Ex. Fr.) Sing. Emir J Food Agric 24(2):113–119

Reyes RG (2000) Indoor cultivation of paddy straw mushroom, *Volvariella volvacea*, in crates. Mycologist 14(4):174–176

Shakil MH, Tasnia M, Munim ZH, Mehedi MHK (2014) Mushroom as a mechanism to alleviate poverty, unemployment and malnutrition. Asian Bus Rev 4(9):31–34

Shwetha VK, Sudha GM (2012) Ameliorative effect of *Volvariella volvacea* aqueous extract (Bulliard ex fries) singer on gentamicin induced renal damage. Int J Pharm Bio Sci 3(3):105–117

Solovyev N, Prakash NT, Bhatia P, Prakash R, Drobyshev E, Michalke B (2018) Selenium-rich mushrooms cultivation on a wheat straw substrate from seleniferous area in Punjab, India. J Trace Elem Med Biol 50:362–366

Sudha A, Lakshmanan P, Kalaiselvan B (2008) Antioxidant properties of paddy straw mushroom (*Volvariella volvacea* (bull. Ex Fr.)) sing. Int J Appl Agric Res 3:9–16

Le Duy Thang (2006). Growing edible mushroom. Ho Chi Minh Agricultural Publisher, 242 p

Thiribhuvanamala G, Krishnamoorthy S, Manoranjitham K, Praksasm V, Krishnan S (2012) Improved techniques to enhance the yield of paddy straw mushroom (*Volvariella volvacea*) for commercial cultivation. Afr J Biotechnol 11(64):12740–12748

Thuc LV, Truc NTT, Hung NV (2019) Outdoor mushroom growing. http://ricestraw.irri.org/resources-1

Tripathy A (2010) Yield evaluation of paddy straw mushrooms (*Volvariella* spp.) on various lignocellulosic wastes. Int J Appl Agric Res 5(3):317–326

USITC (United States International Trade Commission) (2010) Mushrooms industry and trade summary. Office of Industries Publication, Washington, DC

Vivar-Quintana AM, González-San Josè ML, Collado-Fernández M (1999) Influence of canning process on colour weight and grade of mushrooms. Food Chem 66:87–92

Vizhanyo T, Jozsef F (2000) Enhancing colour difference in images of diseased mushrooms. Comput Electron Agric 26:187–198

Walde SG, Velu V, Jyothirmayi T, Math RG (2006) Effects of pretreatment and drying methods on dehydration of mushroom. J Food Eng 74:108–115. https://www.sciencedirect.com/science/article/abs/pii/S0260877405001202

Zhang Y, Geng W, Shen Y, Wang Y, Dai YC (2014) Edible mushroom cultivation for food security and rural development in China: bio-innovation, technological dissemination and marketing. Sustainability 6(5):2961–2973

Zikriyani H, Saskiawan I, Mangunwardoyo W (2018) Utilization of agricultural waste for cultivation of paddy straw mushrooms (*Volvariella volvacea* (bull.) singer 1951). Intl J Agric Technol 14(5):805–814

Chapter 7
Rice Straw-Based Fodder for Ruminants

Daniel Aquino, Arnel Del Barrio, Nguyen Xuan Trach, Nguyen Thanh Hai, Duong Nguyen Khang, Nguyen Tat Toan, and Nguyen Van Hung

Abstract Rice straw is a readily available, practical, and cheap source of fodder for feeding ruminants such as buffaloes, cattle, goats, and sheep. Livestock producers commonly haul and stack rice straw from their rice farm, which then forms reserved feed for their animals during lean months or when good-quality roughages are scarce. The feeding of pure rice straw to ruminants during the stages of fast growth and early lactation has been shown to affect both body condition score and animal performance. This is due to lower dry matter intake and protein content (from 4.0% to 4.7% crude protein) of the straw. The high silica and lignin contents of straw also contribute to poor nutrient (dry matter and protein) digestibility (<50%). So, pre-treatment of straw is necessary to enhance its contribution to improving meat and milk production. Science- and technology-based farm strategies to optimize the nutritive and feeding values of rice straw had been developed with significant improvement on intake, nutrient digestibility, and animal performance. These technologies were also proven effective in contributing additional income to livestock producers from the sales of milk or live animals. This chapter presents and discusses current innovations and developed technologies on how the nutritive (nutrient composition and fiber fraction) and feeding values of rice straw can be improved.

D. Aquino (✉)
Philippine Carabao Center, Central Luzon State University (CLSU), Muñoz, Philippines
e-mail: oed@pcc.gov.ph

A. D. Barrio
Philippine Carabao Center, National Headquarters, Muñoz, Philippines
e-mail: oed@pcc.gov.ph

N. X. Trach
Vietnam National University of Agriculture, Hanoi, Vietnam
e-mail: nxtrach@vnua.edu.vn

N. T. Hai · D. N. Khang · N. T. Toan
Nong Lam University, Ho Chi Minh City, Vietnam
e-mail: hai.nguyenthanh@hcmuaf.edu.vn; toan.nguyentat@hcmuaf.edu.vn

N. V. Hung
International Rice Research Institute (IRRI), Los Baños, Laguna, Philippines
e-mail: hung.nguyen@irri.org

M. Gummert et al. (eds.), *Sustainable Rice Straw Management*,
https://doi.org/10.1007/978-3-030-32373-8_7

Specifically, this focuses on pretreatment (optimization process), enrichment, and recycling of rice straw by physical, chemical, and biological processes. Also covered are practical feeding protocols when rice straw—or its combination with other feed ingredients—is used in formulating a ration. The authors also share secondary information on the effect of rice straw as animal fodder on the improvement in animal performance and production efficiencies; as well as its impact on food production (meat and milk), increasing farmers' income, and on the protection of the environment.

Keywords Rice straw · Fodder · Ruminant · Buffaloes

7.1 Introduction

Livestock farming plays a significant role in agricultural development. Aside from being a source of income for the farmers, livestock also contributes to the production of food for the general public. Ruminant animals, such as buffaloes, cattle, sheep, and goats, are considered economically important in the production of meat and milk, among other derived products such as hides, manure as an organic fertilizer or as fuel/biogas for kitchen use by the livestock-farming families. Ruminants can be entirely dependent on crops for their nourishment to achieve normal growth, production, and reproduction. The dynamics of their rumen ecosystem provides a unique environment for microorganisms to grow and multiply so that these can degrade nutrients, especially fibrous components, from the ingested fodders that eventually are transformed into protein rich foodstuffs such as meat and milk. However, the efficiency of the animal to utilize and convert nutrients from dietary sources into nutritious food products are dependent mainly on the availability and quality of the fodder being offered to the animal.

The availability of quality forages for feeding ruminants is seasonal. Wet season is a time of feed abundance while dry season is a period of scarcity. In countries that experience feed scarcity or deficiency of good quality forages, rice straw remains as the practical, abundant and cheap source of fodder for feeding cattle, buffalo, goat and sheep. According to FAO (2000), the world produced approximately 2000 million tons of cereal straw annually. More than 200 million tons of rice straw was also produced annually in the Southeast Asian countries (see Chap. 1). The estimated quantity of rice straw production is based on the report of Maiorella (1985); and Doyle et al. (1986) that for every hectare of rice farm, the weights of rice grain and rice straw that can be harvested are the same. In many agricultural countries, rice straw and other agro-industrial by-products are available in large quantities immediately every after harvest seasons. These farm byproducts are utilized in many different ways such as fodder for ruminants, for mushroom production, for fuel (heating, biogas) source, for board or paper production and also for organic fertilizer production.

7.2 Rice Straw as a Feed Source

7.2.1 *Availability and Carrying Capacity*

Rice straw is abundantly produced by rice farmers in many agricultural countries worldwide. In Southeast Asia, about 30–40% of the total rice straw production is commonly used to feed more than 90% of the ruminant population in the region including other countries such as China and Mongolia, (Devendra and Thomas 2002). Rice straw is lean-month stuff to ruminant when supply of good quality forages is inadequate. It can be fed as the sole diet to meet the dry matter requirement but it is not a guarantee that other essential nutrients needed for normal body functions by the animal are met. Rice straw can also be offered, up to 60%, in combinations with other feed ingredients, such as concentrates, molasses, or legumes to improve palatability, protein content, and intake and digestibility by the animals.

Every after rice harvest, livestock farmers collect and stock-pile rice straw in a simple shed usually made from locally available materials or stored in piles outdoors. The conserved straw is normally used as animal fodder during the lean months of January through May or when the paddies are already planted with rice in July and August. The extent of rice straw utilization as fodder is dictated by the availability of forage gardens as well as the number of animals being fed. Since the average landholding of the average crop farmer is 1–2 ha, the expected rice straw production annually is only about from 10 to 15 tons, which can support only three or four animal units whether cattle or buffalo. If a farmer has more animals, additional fodder should be sourced out, such as silage, corn stover or hay, or other feed supplements, such as concentrates or legumes, to achieve normal animal performance.

7.2.2 *Nutrients in Rice Straw*

Basically, rice straw has low protein content ranging from 3% to 6%. Is has high cell walls, the neutral detergent fiber (ADF) and acid detergent fiber (ADF) which consisted of the degradable carbohydrate fractions such as starch, cellulose and hemicellulose. It also contains an indigestible phenolic substance called lignin. When used as fodder, rice straw primarily serves as bulk or filler to meet the dry matter requirement of ruminants. This contains 80% substances which are potentially degradable and a source of energy. It has high dry matter (DM) contents of 92–96% but with a low CP content ranging from 3% to 7% (Shen et al. 1998). The lignin and silica contents provide structure to the rice plant during the growing and fruiting stage but these components are in an indigestible form when ingested by animals.

As fodder, rice straw has low energy and protein contents. Its utilization is limited due to minimal contents of digestible nutrients and various characteristics such as palatability, variable nutritional values, high silica and oxalates, and sometimes presence of adulterants when not properly collected and stored. Nevertheless, rice straw still remains to be a practical fodder particularly in times of El Nino or in times of critical periods when sources of fresh fodders are insufficient. In addition, according to a dairy cow nutrient expert from Israel (Hanan Saggi, Feeding & Nutrition Director, TH True Milk Group), rice straw is a good feed component if pretreated properly, particularly when the milk cows are not producing fully.

Rice straw contains higher quantities of potassium (1.58% of DM), calcium (0.53%), and magnesium (0.24%). But it is low in phosphorus (0.12%), sodium (0.13%), iron (0.07%), and manganese (0.07%), (Shen et al. 1998). The bioavailability of these minerals is still to be investigated since most of these minerals are cross-linked to other substances in rice straw in the form of acid-insoluble ash.

The phosphorus (0.02–0.16%) content of rice straw is not sufficient to meet the required 0.3% for growth and normal fertility of animals (Jackson 1977). However, its calcium content of 0.4% is considered adequate to meet the daily requirement for livestock but this does not always hold true. The bioavailability of calcium from rice straw is important to consider since the report of Nath et al. (1969) showed that cattle fed with rice straw has a negative calcium balance even though the calcium content of the straw used in the feeding experiment was apparently adequate. In similar experiments by Joshi and Talapatra (1968), they have higher positive calcium balances with animals fed with wheat straw and sorghum stover diets than on rice straw diets, even though the calcium intake on rice straw diets was higher. According to the authors, when feeding rice straw, it is safe to provide calcium supplementation to the animals.

7.2.3 Rice Straw Intake by Ruminants

Generally, the quantity of rice straw that the animal can eat each day is limited to less than 2% of its body weight. According to the report of Devendra (1997), the amount of rice straw that ruminants can consume can be as high as 1.2 kg DM 100 kg^{-1} of live weight day^{-1}. The rice straw intake, however, varies among animals and this is also influenced by the proportion or parts of the rice straw used in the ration. The intake of rice straw also varies according to the manner in which it is prepared, processed, and fed to the animals. Physical processing, such as chopping or the use of chemical or microbiological treatments, considerably improves an animal's rice straw intake. When offered as is, rice straw intake is lower because it is bulky or occupies more space in the rumen. The digestibility of the straw is also affected due to the slow passage rate of ingested straw and its fermentation by microorganisms in the rumen. Chopping the straw provides more space in the rumen and allows more entries of microorganisms to ferment the straw's degradable components.

7.2.4 Nutrient Digestibility of Rice Straw

The leaf and stem ratio is essential when it comes to the digestibility of cereal straw. Relatively, rice straw has a higher proportion of leaves at 60% compared with other cereal straw, such as barley (35%) and oats (43%) (Sarnklong et al. 2010; Theander and Aman 1984). Having this high proportion of leaves to stems promotes lower in vitro dry matter digestibility (IVDMD) of the leaves at 50–51% compared to the stems at 61% (Vadiveloo 2000). These data were supported by Phang and Vadiveloo (1992) who observed that, in goats, IVDMD for rice leaves is 56.2% while for stems is at 68.5%. To increase the degradability of rice straw leaves, pretreating them with 4% urea solution for 21 days shows significant increase in the IVDMD of the leaves compared with the stems (Vadiveloo 2000). This improvement of the feeding value of rice straw should be taken into consideration to optimize digestibility.

Rice straw, when offered to ruminants, gave DM digestibility ranging from 45 to 50%. Various enzymes secreted in the reiculo-rumen, such as glucanase, cellulase, and hemi-cellulase—not including ligninase—have the potential capacity to degrade the cell wall components of rice straw, (Schiere and Ibrahim 1989). These enzymes are not produced by the animals themselves but are secreted by the rumen microorganisms. The degree of lignification or with higher the lignin content, this has a direct effect on the reduction of the rice straw's nutrient digestibility. In addition, Agbagla-Dohnani et al. (2003) pointed out that silica has a direct effect on cell wall digestibility of rice straw since silica forms a physical barrier that lowers microbial degradation resulting to poor enzymatic hydrolysis of the straw.

7.3 Pretreatment of Rice Straw as Ruminant Fodder

Developed technologies have been published and are available for farmers to help them enhance the utilization and improvement of the nutritive value of rice straw for animal feeding. These techniques include different physical, chemical, and biological processing methods and combinations of these (Ibrahim 1983). However, adoption of these technologies takes time since they require additional inputs and farmers need to see improvement to believe it.

7.3.1 Physical Processes

The physical process is a practical and inexpensive method to enhance utilization and recycling of nutrients from rice straw when used as fodder for ruminants. Physical treatment of rice straw aims to improve the palatability and increase intake as well as improve the potential digestibility of ruminants. These physical processes include soaking, grinding or chopping, pelleting, steaming pressure, and gamma irradiation. These processes promote physical changes in rice straw, such as reducing particle size, which lessens rumination time for the animal; enriching softness of the straw's fibrous components to make it more palatable to the animal; and hastening nutrient digestion.

Soaking is a common and economical process of treating rice straw. This is being done by soaking straw overnight in water which brings softness between the of lignin and cellulose component of rice straw. Soaking of straw promotes higher intake of the animal as well as nutrients digestibility. Soaking along with steaming technique have direct effect on the cell walls delignification of rice straw, (Walker 1984). The effect of steam or exposure of the lignocellulosic contents of rice straw under high pressure provides a good environment for the microbial enzymes for faster fermentation of nutrients, thus increasing the rice straw digestibility (Walker 1984). Milstein et al. (1987) suggested that heat treatment leads to an increase in cellulose digestibility from 20% to 40%.

Grinding, chopping or pelleting had beneficial effects in breaking down the cell wall contents of rice straw. These physical processes reduced the particle size of the straw thus, providing easy entries or access of the rumen microorganisms for degradation. The use of these techniques should properly consider the balance between the particle size and the retention time or passage rate of the ingested treated straw. The reduction in particles due to grinding or chopping of rice straw promotes animal intake and increase passage rate of the feed, however, this brings negative effect in terms of decreasing the nutrients digestibility of straw. This is because of the less time exposure of the feed materials for rumination and for microbial fermentation in the rumen.

Pressure steaming rice straw is another process to consider. However, the process may add cost for farmers due to the energy required during process. Rangnekar et al. (1982) and Liu et al. (1999) have tried steam treatment under high pressure of 15 bar for 5 min at a moisture level varying between 30% and 70% (w/w) using different roughages and rice straw. They observed that the different fractions of rice straw, such as hemicellulose, cellulose, lignin, and sugars were separated by steam pressure. Similar observations were also reported by Ooshima et al. (1984) when irradiated rice straw was subjected to 84% water content in microwaves (2450 MHz) using sealed glass vessels with accessible partitions into cellulosic materials and with increase digestible nutrients of the straw.

7.3.2 Chemical Treatment

The chemical method to improve the nutritive value of rice straw has been done for more than 100 years (Kamstra et al. 1958) with the aim to increase animals' intake and feed digestibility. The chemicals, which are commonly studied and used in treating rice straw to improve its palatability, intake, and digestibility, are sodium hydroxide, ammonia, and urea. The mode of action of these chemicals is to break the links between the lignin-cellulose structures of the straw, which are sensitive under alkaline or acidic conditions. Among the chemicals used, alkali agents are extensively explored and practically accepted under farm conditions. During straw treatment, basic chemicals, such as sodium hydroxide, urea, or ammonia are absorbed into the cell wall and react with the lingo-cellulosic contents of the straw to break the ester bonds between lignin and hemicellulose and cellulose. The alkali absorbed into the straw directly causes the structural fibers to swell making it free for microbial fermentation (Chenost and Kayouli 1997; Lam et al. 2001).

7.3.2.1 Sodium Hydroxide (NaOH) Treatment

The use of NaOH in the treatment of cereal straw has been done since the 1940s (Mcanally 1942). The straw is treated using 1.5% NAOH w/w for 24 h in a container. The treated straw is rinsed with cold water and subjected to in vitro digestibility. Results showed that the NAOH treated straw is more digestible than the pure straw by as much as 28%. The Beckman method, which is similar to the procedures of Mcanally (1942) for the NAOH treatment of straw, has been recommended by FAO (2012). The Beckman method also uses 1.5% NAOH but the treatment period is within 18–20 h before rinsing with tap water. The NAOH acts on the straw by reducing proteolysis and increasing delignification by unlocking the linkage between the lignin and cellulosic contents of the straw to give more time for microbial enzymatic action to take place. Treatment of rice straw and other crop residues using NAOH has been reviewed by Jackson (1977), Berger et al. (1994), Arieli (1997), and Wang et al. (2004). These authors concluded that chemical reactions of NAOH on the cell wall contents of rice straw is advantageous for the breakdown of the esterified bonds between the phenols group and the cellulosic components of straw thus favoring the enzymatic hydrolysis.

Feeding NAOH-treated straw in cattle showed better performance than ammonia treatment of straw. Similar improvement in animal performance was also reported by Chaudhry and Miller (1996) and Vadiveloo (2000) when NAOH-treated rice straw was fed to cattle compared to untreated straw. This was due to the improvement in palatability and intake of the animals and increase in digestibility of treated straw. The adoption of NaOH treatment of rice straw, however; is not widely practiced by farmers. This is because NAOH costs more than urea treatment and it is not always available. In addition, NaOH, when used at higher concentrations, poses health problems for animals if the amount exceeds 10 g of the daily sodium requirement of mature animals. It can also cause pollution problems due to sodium accumulation in the environment (Sundstol and Coxworth 1984).

7.3.2.2 Ammonia (NH_3) Treatment of Rice Straw

Treating rice straw using anhydrous and aqueous ammonia, urea, and other ammonia-releasing substances have been investigated and have been proven to enhance the degradability of the straw (Abou-EL-Enin et al. 1999; Selim et al. 2004;

Fadel-Elseed et al. 2003). The treatment of rice straw with ammonia (NH_3) is similar to treating with NaOH. NH_3 has been observed to be advantageous over the use of NAOH because it is readily available because it can be derived from the hydrolysis of urea. NH_3 treatment does not only increase degradability of rice straw but it also supplies nitrogen (Abou-EL-Enin et al. 1999), thereby increasing the protein content of the straw. It can also be used as a preservative agent since it inhibits the growth of molds in the treated straw (Calzado and Rolz 1990). Other benefits that can be derived from NH_3 treatment include reducing costs of buying protein-rich supplements and enhancing acceptability and voluntary intake of the treated straw by ruminants.

Liu et al. (2002) observed that the use of NAOH treatment is more efficient than NH_3 treatment in terms of improving the energy values of the straw. However; using NH_3 is usually more profitable for farmers than NAOH because it provides an additional source of nitrogen in the straw. Selim et al. (2004) studied sheep fed with NH_3-treated rice straw packed in polyethylene bags for 4 weeks with gaseous ammonia (3 g NH_3 100 g dry matter^{-1}). NH_3 increased the N content of the treated rice straw from 8.16 to 18.4 g kg^{-1} or with an equivalent increase of CP from 51 to 115 g kg^{-1}. A slight decrease in the NDF of treated straw (from 571 to 551 g kg^{-1}) was observed but with an increase in acid detergent fiber (ADF) from 303 to 327 g kg^{-1}. This further indicated positive changes on the cell wall content of the treated straw.

7.3.2.3 Urea Treatment

Urea treatment is the most practical and widely used chemical method in treating rice straw. It is adoptable by both small-scale and commercial livestock farms. The main function of urea is to increase the protein content of the treated straw during the fermentation process. Urea or NH_3 is best used in combination with molasses (urea-molasses solution) at 30% moisture content of the treated straw. First, urea is hydrolyzed or undergoes ureolysis to produce ammonia-nitrogen (Sahnounea et al. 1991). The role of the molasses is to supply energy so that cellulosic fermentation of the treated straw is hastened. Urea or its combination with molasses can make rice straw a complete and safe basal ration for ruminants (Langar et al. 1985).

Rice straw can be effectively treated with urea using different concentrations i. e. from 1% to 5% w/w. Urea should be dissolved first in water at the desired proportion and it can be sprayed into the rice straw. The treated straw can be packed in the silo, empty drum or plastic bag. This treatment process is practical and can be easily adopted by farmers. Urea is a chemical which is a source of nitrogen to crops and a source of non-protein nitrogen to ruminants. It is a crystalline substance and it is easy to handle and locally available in the market, (Sundstøl and Coxworth 1984). Urea increases the nitrogen (crude protein or CP) content of the treated rice straw, (Schiere and Ibrahim 1989). It is cheaper than NaOH or pure NH_3. Vadiveloo (2003) reported that treating different rice varieties with low degradable carbohydrates responded positively compared to high-quality rice straw varieties after urea treatments as reflected by the increase in IVDMD from 45% to 55–62%. Numerous evaluations were done in the laboratory (Reddy 1996; Shen et al. 1998; Vadiveloo 2003) or in field trials (Prasad et al. 1998; Vu et al. 1999; Akter et al. 2004) in treatment of rice straw using pure urea or in combination with other chemicals or feed supplements and the results had clear improvement on the nutritive as well as feeding value of treated straw.

7.3.2.4 Lime Treatment

Treatment of straw with lime solution [$CaO/Ca(OH)_2$] is expected to have the same effect on improving fiber degradability as NAOH. Lime is also a source of calcium for ruminants in low-calcium rations but it has longer solubility in water compared to NAOH or urea. Treatment of straw with lime can be done in two ways: by soaking and ensiling. Lime treatment provides complementary effects in combination with urea. The combination of lime and urea has been shown an advantage in increasing degradability and incrementing both the calcium and nitrogen contents of the treated straw (Nguyen 2000).

In a separate study of Pradhan et al. (1997), using 4% or 6% $Ca(OH)_2$ to treat rice straw, showed, after ensiling, a higher IVDMD. However; it is further suggested that a combination of lime and urea would give better results than either urea or lime alone. Sirohi and Rai (1995) used 3% urea plus 4% lime at 50% moisture for 3 weeks of incubation. They found this to be the most effective treatment process for rice straw. This was due improving the digestibility and degradable nutrients of the treated straw. Saadulah et al. (1981) and Hadjipanayiotou (1984) found that the use of lime and other alkali agents had additive effects on rice straw treatment and utilization in addition to being safer and more cost-effective to use than NaOH.

As cited by Trach et al. (2001), there are reports that treated rice straw with pure lime posed contradicting results in its effect on delignification or degradation of rice straw. There was a report that the dry matter intake of animals was reduced due a palatability problem of the treated straw. Lime treatment did not affect N content, but it appeared to be more powerful in delignification or reducing neutral detergent fiber (NDF) and hemicellulose contents of the treated straw. Increasing levels of lime and/or urea during rice straw treatment resulted in some negative interactions between the two chemicals. However, a level of 2% urea alone seemed to be too low for effective treatment and a level of 6% lime seemed to be too high for rumen cellulolysis.

7.3.3 Biological Treatment

Biological treatment of rice straw involves the use of enzymes and different microorganisms, such as bacteria and fungi. Different fungi strains have the capacity to act on the cell wall contents of the straw thereby improving the degradation rates and making other nutrients available to the animal. As cited by Jalc (2002), the enzymes secreted by fungi had strong affinity to metabolize lingo-celluloses and these are biological agents in treating rice straw to improve its nutritional value through the selective action of delignification. Nevertheless, its current use in developing countries is still a big question due to limitation in technical skills and the availability of resources to produce and handle large quantities of fungi or their enzymes for practical and field application. Biological treatment of straw brings some concerns and problems to be addressed and overcome (Schiere and Ibrahim 1989). For example, there are fungi species that are not edible and produce toxic substances both to human and animals. Fungi also require an environment for them to grow and reproduce, such as pH, temperature, pressure, and O_2, and CO_2 concentrations before, during, and after the treatment period. With the current development in mycology, there are now simple protocols or guides to be used in growing fungi as well in enzyme production or purification for rice straw treatment. There are commercially available enzyme inoculants or additives available in the market such that the costs to purchase these substrates will continuously decline and can be used by ruminant raisers to increase their production efficiency as well as their farm income (Beauchemin et al. 2004).

7.3.3.1 White-Rot Fungi Treatment

White-rot fungi are known to have degrading or decaying properties by acting ligno-cellulolytic components of farm byproducts including wood. These have the capacity to decompose and metabolize cellulose, hemicellulose, and lignin under favorable environments through enzymatic reactions to their substrates (Eriksson et al. 1990). Some of the significant characteristics of many white-rot fungi species involve their ability to effectively hydrolyze lignin hence they are considered to be lignin degraders. These species can improve the nutritive value of fodder by tendering more degradable carbohydrates for rumen microbial fermentation (Yamakava and Okamnto 1992; Howard et al. 2003). White-rot fungi secrete varieties of extracellular lignin-modifying enzymes that consist of lignin-peroxidase (LiP), manganese-dependent peroxidase (MnP), laccase (phenol oxidase), and H_2O_2-producing oxidase (aryl-alcohol oxidase; AAO and glyoxaloxidase) (Kirk and Farrell 1987; Arora et al. 2002; Novotny et al. 2004; Arora and Gill 2005; Lechner and Papinutti 2006).

Researchers have observed that some fungi species can decompose or directly act on free phenolic monomers to break the bonds or cross-links between lignin and polysaccharides of rice straw (Chen et al. 1996). Other fungal species improve the IVDMD of treated straw (Karunanandaa et al. 1995; Karunanadaa and Varga 1996a, b; Fazaeli et al. 2006). Karunanandaa et al. (1995) also reported that incubation of rice straw with 8–10% w/w for 30 days using three white-rot fungi species. *Pleurotus sajor-caju* enhanced IVDMD in both rice leaves and stems. However, results obtained using *Cyathus stercoreus* gave the highest IVDMD compared to other fungi species (Karunanandaa et al. 1992). The sequence by which the white-rot fungi act on its substrates is dependent on the fungal species. There are species that prefer to access first on readily degradable carbohydrates, such as simple sugars, cellulose, and hemicellulose and eventually degrade lignin, thus resulting in a lower energy supply for ruminants (Karunanadaa and Varga 1996a, b; Jalc 2002). The length of incubation in treatment of straw is dependent on the white-rot fungi species. During the early stage of incubation, some losses in energy are expected due to mycelial growth but after a certain time, some white-rot species preferably attack lignin without degrading cellulose and hemicellulose, thus supplying more degradable energy for the ruminants.

Nowadays, it is important to do research on mycology by selecting fungi species that prefer to attack lignin rather than the structural carbohydrates or cell walls of rice straw. Once these species are identified, mycologists can breed even better strains (Rodrigues et al. 2008). Growing edible mushrooms is a dual purpose of treating rice straw. As described elsewhere in this book, rice straw serves as a substrate to produce food (mushrooms) and feed from the mushroom-spent bedding. Some of the edible fungal species include *Pleurotis ostreatus* and *Volvarella* sp. These can be grown easily and the left-over mycelia from the mushroom bedding can increase the protein as well as the degradable carbohydrates of the rice straw. Continuous research on white-rot species has to be done and identification of new edible fungi species is necessary to explore the potential and characteristics to produce more fruiting bodies for farmers' harvest as well as achieving optimum feeding quality of the unutilized mushroom bedding.

7.3.3.2 Treatment with Enzymes

The catabolic breakdown of any complex substance into its simplest component is brought about by chemical reactions and/or by enzymatic processes. Enzymes involved in the degradation of rice straw are mostly of microbial origin and their action is very specific to the substrates to be degraded. There are commercially-available fiber-degrading enzymes, such as cellulases, hemicellosi, glucanase, and xylanases and many others. However, their stability and potency are always affected by many factors, such as temperature and duration as well as how the enzyme products were processed and packaged. Commercial enzymes used in the livestock feed industry are generally of fungal (*Trichoderma longibrachiatum*, *Aspergillus niger*, and *A. oryzae*) or bacterial (*Lactobacillus* and *Staphylococcus species*) origins (Colombatto et al. 2003).

The degradability of cereal straw can be increased through enzyme treatment or any combination of other treatments (Liu and Ørskov 2000; Wang et al. 2004; Zhu et al. 2005; Eun et al. 2006; Fazaeli et al. 2006; Rodrigues et al. 2008). Additionally, using fibrolytic enzymes show improvements in the average daily gain of steers (Beauchemin et al. 1995), fleece weight and wool production of lambs (Jafari et al. 2005), and milk production of dairy cows (Yang et al. 2000). Enzyme treatment of rice straw is not yet very popular in raising ruminants under small-scale production systems because of the additional input costs involve as well as the limitation of skills for using enzyme products.

7.4 Effects of Feeding Pure or Pretreated Rice Straw to Ruminants

Generally, in feeding dry cows, rice straw can be used for about 50% of the ration. Additional urea-molasses mineral blocks could be used as supplements to support the requirement of the dry cows. Rations with rice straw greater than 50% would result in a declining body weight of the cows.

For cows with calves, the use of rice straw should not exceed 25% of the total ration, with the remaining 75% being good-quality hay or legumes or a concentrate supplement. When feeding lactating cows, rice straw alone is not adequate to support milk synthesis or milk production. Supplementary feeds, such as dairy concentrates or dried legumes, are required to augment the deficient nutrients in rice straw so that the goal of supporting normal milk production is achieved.

One consideration in feeding rice straw to ruminants is to balance the quantity of phosphorus and other trace minerals in the ration. Rice straw has lower phosphorus and trace mineral contents, thus supplementation with trace minerals and phosphorus, especially in high-yielding cows, is necessary.

7.4.1 Effects of Urea-Treated Rice Straw in Ruminants

Aquino et al. (2016) reported on the effects of feeding urea-molasses-treated rice straw to dairy buffaloes through the community science and technology-based farm project involving 30 dairy buffalo farmers in the Philippines. The farmers were trained to produce treated rice straw using urea-molasses solution (UMS). The UMS consisted of 2% urea, 5% molasses, and 93% water at a 2-parts rice straw to 1-part UMS ratio. The treated rice straw was allowed to partially ferment in silage bags for 21 days before feeding to buffaloes. Results of feeding UMS-treated rice straw (UMTRS) to dairy buffaloes showed a total milk production of 974 kg cow^{-1} in 210 milking days. In contrast, buffaloes fed no UMTRS produced 777 kg of milk during the same lactation period. Comparing the effect of UMTRS feeding with that of pure rice straw showed a difference of 147 kg milk production or with a milk yield difference of 0.7 kg milk cow^{-1} day^{-1}. It was also noted that the UMS improves the crude protein content of treated rice straw from 4.7% to 7.9% and the DM digestibility of rice straw was increased from 47% to 55%.

In a separate study, Aquino et al. (2018) used fermented total-mixed rations (FTMRs) composed of rice straw (RS) in combinations with banana byproducts or water hyacinth (Table 7.1). The formulated FTMRs composed of other feed

Table 7.1 ADG in weight of growing and lactating buffaloes fed fermented total mixed ration composed of rice straw and banana byproducts or water hyacinth

Item	Control diet	Rice straw + banana byproducts		Rice straw + water hyacinth	
		25%	50%	25%	50%
Growing buffaloes, #	**5**	**5**	**5**	**5**	**5**
Initial weight, kg	218.60	220.60	222.00	247.80	247.80
Final weight, kg	293.60	282.20	304.40	315.00	268.40
ADG, kg	**0.81**	**0.68**	**0.96**	**0.67**	**0.28**
Lactating buffaloes, #	**5**	**5**	**5**	**5**	**5**
Total milk yield, kg $120d^{-1}$	800.40	752.40	764.00	812.00	793.20
Milk yield, kg d^{-1}	**6.67**	**6.27**	**6.37**	**6.77**	**6.61**

ingredients, such as rice bran, copra meal, molasses, mono di-calcium phosphate, and urea. The FTRMs had remarkable results in terms of ADG and milk production of dairy buffaloes. The FTMR, composed of 20% rice straw in combination with 50% banana byproducts, resulted in a 960-g ADG compared to a 810-g ADG of growing buffaloes in the control ration. This brought an 18.9% increase in the growth rate of the buffaloes. On the other hand, FTMR, composed of 28% rice straw combined with 25% water hyacinth, gave a 670-g ADG compared to only a 520-g ADG for the control diet, which is equivalent to an increase of 28.85%. The increase in ADG of growing buffaloes was attributed to the increase in daily feed intake from 1.7% (control) to 2.13% (50% banana + RS) and from 2.01% (control) to 2.65% (25% water lily +RS) of the body weight. In addition, there was an increase in DM digestibility (from 50.95% to 60.35%) and CP digestibility (62.30–66.33%) for rice straw with 50% banana byproducts. The combination of RS with 25% water lily also improved the DM (50.10% vs. 57.96%) and CP digestibility (58.08% vs 61.96%), respectively.

The FTMR with 28% rice straw plus 25% water hyacinth was recommended over the control diet as shown by a 100-g milk difference over the control (6.77 vs 6.67 kg day^{-1}) or FTMR with 50% banana byproducts with a 400-g milk difference (6.77 vs 6.37 kg) over the FTMR with 50% banana byproducts. The observations were also supported by the increase in the daily DM intake; 2.5% vs 2.3% of body weight of the cows.

The ration composed of rice straw with supplementary protein, energy, and/or minerals have been shown to optimize rumen function and maximize the utilization or intake of rice straw. Chenost and Kayouli (1997) emphasized that rumen microorganisms should be provided with needed nutrients for their growth and self-multiplication so that degradation of the cell walls of straw is maximized. This also leads to conditions for sustainable process of cellulolysis. In a field trial, Warly et al. (1992) showed that a rice straw ration with supplementary soybean meal increased both degradability and intake of the animals. Untreated rice straw is low in protein when this is supplemented with cottonseed meal (Wanapat et al. 1996) or urea molasses-multi-nutrient block (Vu et al. 1999; Wanapat et al. 1999; Akter et al. 2004); these significantly increase the cow's milk production.

7.4.2 *Effects of Biological Treatment of Rice Straw*

Zadrazil (1977) identified three species of fungi based on substrate preference and type of enzymes they secrete for the degradation rice straw cell walls. The first group has cellulolytic and hemicellulolytic activities of which they act on cellulose and hemicellulose. The second group of fungi preferentially acts on the lignin content while the third group of fungi decomposes cellulose, hemicellulose, and lignin simultaneously. The second group of fungi is the most recommended for rice straw treatment because of its peculiarity to break and degrade structural carbohydrates present in rice straw. It is suggested that screening new fungal strains is essential with desired characteristics to efficiently improve the nutritive and feeding value of rice straw.

Zayed (2018) evaluated different parameters for the improvement of the nutritional value of rice straw. During his evaluation, he used moist straw, soaked straw for 24 h without pasteurization, and soaked straw for 24 h with pasteurization at 100 °C for 1 h. The preprocessed rice straw samples were inoculated having three combinations of microbial inoculants. He also observed that moistened rice straw had the highest organic matter reduction at 74.21% if inoculated with *Azotobacter chroococcum* and *Saccharomyces cerevisiae*. Additionally, if inoculated with *Azospirillum brasilense* and *Saccharomyces cerevisiae,* significant reduction in crude fiber at 27.54%; neutral detergent fiber at 55.39%; and 42.47% acid detergent fiber can be observed. For rice straw soaked for 24 h and inoculated with *Azospirillum brasilense* and *Bacillus megaterium,* a significant increase in crude protein at 13.71% was observed. Zayed (2018) further concluded that interaction between microbial treatment and physical pretreatments of rice straw shows a significant decrease in organic matter, crude fiber, neutral detergent fiber, and acid detergent fiber as well as a significant increase in crude protein compared to the control.

7.5 Limitations of Rice Straw Utilization

Several factors were identified that limit the utilization of straw as animal fodder. These include poor digestibility, low animal intake, and very low protein content. Technologies to overcome the identified factors have been developed for pretreatment of straw before feeding to animals. However, its adoptability varies according to the capacity and capability of the farmers or its practicality including health and environmental concerns when used by the farmers.

In physical treatment of straw, the limitation is mainly on grinding of the straw into smaller particle size. The positive effect of reduced particle size is that it promotes higher intake due to an increase in the rate of passage of the ingested feed by the animal. The negative side of this is that it causes less time for rumination and less exposure to microbial degradation, thus in turn reducing degradation and digestibility of the straw components. Uden (1988) observed that grinding and pelleting of grass hay decreased dry matter degradability in cows from 73% to 67%, which was mainly due to a decreased fermentation rate (9.4–5.1% h^{-1}) and decreased total retention time of the solids from 73 to 54 h, resulting in an increased intake (Stensig et al. 1994). The use of machines in physical treatment and processing of crop residues is also not practical for small-scale farms because of their capacity to buy equipment and the benefits derived may be too low or even negative for the farmers (Schiere and Ibrahim 1989).

The costs involve can be one of the factors that limit the adoption of chemical treatment of straw. Although there are significant effects on the improvement of the nutritive value, animal performance, as well as an increase in income due to treatment of rice straw, the farmers should still balance their decision whether to adopt or not to adopt using treated straw. Hazard issues, such as toxicity and environmental pollution, are some of the limitations in using chemicals for straw treatment.

For microbial treatment of rice straw, one of the major drawbacks is the strain of the fungi to be used and its capacity to degrade lignin and other components of the straw, such as cellulose and hemi-cellulose. Incubation period is another limitation for its practical application in treating straw. There are species of fungi with very high affinity to degrade lingo-cellulosic materials, even in just 1 or 2 weeks of incubation and these have to be explored for its optimum incubation time to increase the feeding value of straw. In addition, some fungi produce toxins that may affect both human and animals so proper care should be considered in using these for rice straw treatment.

7.6 Summary and Recommendations

Among agricultural byproducts, rice straw is most abundant, low in cost, and a practical source of fodder for ruminants. Its utilization as a livestock feed is limited due to problems in collection, hauling, and storage. Rice straw has low nutritional value (low protein content and poor digestibility) compared to grasses, thus it cannot support the nutrients required by high-yielding milk cows and buffaloes. There are technologies that have been developed to increase the nutritive value, nutrient digestibility, and utilization of rice straw, such as physical processing, pretreatment using chemicals, and/or biological treatment. However, adoption of these developed technologies is still low due to farmers' limited skills and inputs (e.g., farm equipment) and their doubts regarding applicability to the farm situation and the benefits for the animals and livestock producers. To maximize the utilization of rice straw as fodder for ruminants, mechanization is most important to facilitate collection, hauling, and stacking or processing all available rice straw from the field.

References

Abou-EL-Enin OH, Fadel JG, Mackill DJ (1999) Differences in chemical composition and fibre digestion of rice straw with, without, anhydrous ammonia from 53 rice varieties. Anim Feed Sci Technol 79:129–136

Agbagla-Dohnani A, Noziere P, Gaillard-Martinie B, Puard M. Doreau M (2003) Effect of silica content on rice straw ruminal degradation. J Anim Sci 140:183–192

Akter Y, Akbar MA, Shahjalal M, Ahmed TU (2004) Effect of urea molasses multi-nutrient blocks supplementation of dairy cows fed rice straw and green grasses on milk yield, composition, live weight gain of cows and calves and feed intake. Pak J Biol Sci 7(9):1523–1525

Aquino DL, Fujihara T, Baltazar H, Santos J (2016) Community-based S & T farm project on the preparation and utilization of urea-treated rice straw (UTRS) as fodder for dairy buffaloes. Proc. PCC R & D review of completed and on-going research projects. July 27–29, 2016

Aquino DL, Casamis, Amido R, Florendo PC, Garcia NP, Baltazar C, Del Barrio AN (2018) Nutritive value, digestibility and performance of buffaloes using banana by-products and water hyacinth as alternative feed sources. 2018 PCC R & D Review of completed and on-going research projects, PCC-NHQ. July 4–6, 2018

Arieli A (1997) Whole cottonseed in dairy cattle feeding: a review. Anim Feed Sci Technol 72:97–110

Arora DS, Gill PK (2005) Production of ligninolytic enzymes by *Phlebia floridensis*. World J Microbiol Biotechnol 21:1021–1028

Arora DS, Chander M, Gill PK (2002) Involvement of lignin peroxidase, manganese peroxidase and laccase in degradation and selective ligninolysis of wheat straw. Int Biodeterior Biodegradation 50:115–120

Beauchemin KA, Rode LM, Sewalt VJH (1995) Fibrolytic enzymes increase fiber digestibility and growth rate of steers fed dry forages. Can J Anim Sci 75:641–644

Beauchemin KA, Colombatto D, Morgavi DP (2004) A rationale for the development of feed enzyme products for ruminants. Can J Anim Sci 84:23–36

Berger LL, Fahey GC, Bourquin LD, Tilgeyer EC (1994) Modification of forage quality after harvest. In: Fahey C (ed) Forage quality, evaluation, and utilisation. American Society of Agronomy, Inc, Madison, pp 922–966

Calzado JF, Rolz C (1990) Estimation of the growth rate of *Pleurotus* on stocked straw. J Ferment Bioeng 69:70–71

Chaudhry AS, Miller EL (1996) The effect of sodium hydroxide and alkaline hydrogen peroxide on chemical composition of wheat straw and voluntary intake, growth and digesta kinetics in store lambs. Anim Feed Sci Technol 60:69–86

Chen J, Fales SL, Varga GA, Royse DJ (1996) Biodegradability of free monomeric cell-wall-bound phenolic acids in maize Stover by two strains of white rot fungi. J Sci Food Agric 71:145–150

Chenost M, Kayouli C (1997) Roughage utilisation in warm climates. FAO animal production and health paper 135, Rome

Colombatto D, Morgavi DP, Furtado AF, Beauchemin KA (2003) Screening of exogenous enzymes for ruminant diets: relationship between biochemical characteristics and in vitro ruminal degradation. J Anim Sci 81:2628–2638

Devendra C (1997) Crop residues for feeding animals in Asia: technology development and adoption in crop/livestock systems. In: Renard C (ed) Crop residuals in sustainable mixed crop/livestock farming system. CAB International, Wallingford, pp 241–267

Devendra C, Thomas D (2002) Crop-animal interactions in mixed farming systems in Asia. Agric Syst 71:27–40

Doyle PT, Devendra C, Pearce GR (1986) Rice straw as a feed for ruminants. International development program. Australia Universities and Colleges, Canberra. https://idl-bnc-idrc.dspacedirect.org/bitstream/handle/10625/8776/70328.pdf?sequence=3

Eriksson K-EL, Blanchette RA, Ander P (1990) Microbial and enzymatic degradation of wood and wood components. Springer, Berlin

Eun JS, Beauchemin KA, Hong SH, Bauer WM (2006) Exogenous enzymes added to untreated or ammoniated rice straw: effects on in vitro fermentation characteristics and degradability. Anim Feed Sci Technol 131:86–101

Fadel Elseed AMA, Sekine J, Hishinuma M, Hamana K (2003) Effects of ammonia, urea plus calcium hydroxide and animal urine treatments on chemical composition and *in sacco* degradability of rice straw. Asian-Aust J Anim Sci 16:368–373

FAO (Food and Agriculture Organization) (2000) Rice straw as livestock feed. http://www.fao.org/3/X6512E/X6512E07.htm

FAO (Food and Agriculture Organization) (2012) Retrieved March 30, 2012 http://Teca.Fao.Org/Technology/Treating-Straw-Animal-Feeding-Beckmann-Method

Fazaeli H, Azizi A, Amile M (2006) Nutritive value index of treated wheat straw with *Pleurotus* fungi fed to sheep. Pak J Biol Sci 9(13):2444–2449

Hadjipanayiotou M (1984) Effect of level and type of alkali on the digestibility *in vitro* of ensiled, chopped barley straw. Agric Wastes 10:187–194

Howard R, Abotsi E, Jansen EL, Howard S (2003) Lignocellulose biotechnology: issues of bioconversion and enzyme production. Afr J Biotechnol 2:602–619

Ibrahim MNM (1983) Physical, chemical, physico-chemical, and biological treatments of crop residues. In: Pearce GR (ed) The utilization of fibrous agricultural residues. Australian Development Assistance Bureau, Research for Development Seminar 3, Los Baños

Jackson MG (1977) Review article: the alkali treatment of straw. Anim Feed Sci Technol 2:105–130

Jafari A, Edriss MA, Alikhani M, Emtiazi G (2005) Effects of treated wheat straw with exogenous fibre-degrading enzymes on wool characteristics of ewe lambs. Pak J Nutr 4:321–326

Jalc D (2002) Straw enrichment for fodder production by fungi. In: Kempken F (ed) The Mycota XI agricultural applications. Springer, Berlin, pp 19–38

Joshi DC, Talapatra SK (1968) Studies on the utilization of minerals under acid- and alkali-producing cattle feeds. Indian J Vet Sci Anim Husb 38:645–664

Kamstra L, Moxon AL, Bentley OG, (1958) The effect of stage of maturity and lignification on the digestion of cellulose in forage plants by rumen microorganisms in vitro. J Anim Sci:199–208

Karunanadaa K, Varga GA (1996a) Colonization of rice straw by white rot fungi (*Cyathus stercoreus*): effect on ruminal fermentation pattern, nitrogen metabolism, and fiber utilization during continuous culture. Anim Feed Sci Technol 61:1–16

Karunanadaa K, Varga GA (1996b) Colonization of rice straw by white rot fungi: cell wall monosaccharides, phenolic acids, ruminal fermentation characteristics and digestibility of cell wall fiber components in vitro. Anim Feed Sci Technol 63:2732–2788

Karunanandaa K, Fales SL, Varga GA, Royse DJ (1992) Chemical composition and biodegradability of crop residues colonized by white rot fungi. J. Sci. Food Agric. 60:105–112

Karunanandaa K, Varga GA, Akin DE, Rigsby LL, Royse DJ (1995) Botanical fractions of rice straw colonized by white rot fungi: changes in chemical composition and structure. Anim Feed Sci Technol 55:179–199

Kirk TK, Farrell RL (1987) Enzymatic "combustion": the microbial degradation of lignin. Ann Rev Microbiol 41:465–505

Lam TBT, Kadoya K, Iiyama K (2001) Bonding of hydroxycinnamic acids to lignin: ferulic and p-coumaric acids are predominantly linked at the benzyl position of lignin, not the b-position, in grass cell walls. Phytochemistry 57:987–992

Langar RN, Bakshi MRS, Gupta VK (1985) Utilization of agricultural wastes. Proc. Symposium, Punjab Agricultural University, Ludhiana

Lechner BE, Papinutti VL (2006) Production of lignocellulosic enzymes during growth and fruiting of the edible fungus *Lentinus tigrinus* on wheat straw. Process Biochem 41:594–598

Liu JX, Ørskov ER (2000) Cellulase treatment of untreated and steam pre-treated rice straw-effect on *in vitro* fermentation characteristics. Anim Feed Sci Technol 88:189–200

Liu JX, Ørskov ER, Chen XB (1999) Optimization of steam treatment as a method for upgrading rice straw as feeds. Anim Feed Sci Technol 76:345–357

Liu JX, Susenbeth A, Südekum KH (2002) *In vitro* gas production measurements to evaluate interactions between untreated and chemically treated rice straws, grass hay, and mulberry leaves. J Anim Sci 80:517–524

Maiorella BL (1985) Ethanol. In: Young M (ed) Comprehensive biotechnology. Pergamon, Oxford, pp 861–914

Mcanally RA (1942) Digestion of straw by the ruminant. Biochem J:392–399

Milstein O, Bechar A, Shragina L, Gressall J (1987) Solar pasteurization of straw for nutritional upgrading and as substrate for ligninolytic organisms. Biotechnol Lett 9:269–274

Nath K, Sahai K, Kehar ND (1969) Effect of water washing, lime treatment and lime and calcium carbonate supplementation on the nutritive value of paddy *(oryza sativa)* straw. J Anim Sci 28:383

Nguyen XT (2000) Treatment and supplenetation of rice straw for ruminant feeding in Vietnam. Doctor Scientiarum Thesis 2000:26. Agricultural University of Norway. ISSN 0802-3220. ISBS 82-575-0440-8

Novotny C, Svobodova K, Erbanova P, Cajthaml T, Kasinath A, Lang E, Sasek V (2004) Ligninolytic fungi in bioremediation: extracellular enzyme production and degradation rate. Soil Biol Biochem 36:1545–1551

Ooshima H, Aso K, Harana Y, Yamamota T (1984) Microwave treatment of cellulosic materials for their enzymatic hydrolysis. Biotechnol Lett 6:289–294

Phang OC, Vadiveloo J (1992) Effects of varieties, botanical fractions and supplements of palm oil by-products on the feeding value of rice straw in goats. Small Rumin Res 6:295–301

Pradhan R, Tobioka H, Tasaki I (1997) Effect of moisture content and different levels of additives on chemical composition and in vitro dry matter digestibility of rice straw. Anim Feed Sci Technol 68:273–284

Prasad RDD, Reddy MR, Reddy GVN (1998) Effect of feeding baled and stacked urea treated rice straw on the performance of crossbred cows. Anim Feed Sci Technol 73:347–352

Rangnekar DV, Badve VC, Kharat ST, Sobale DN, Joshi AL (1982) Effect of high pressure steam treatment on chemical composition and digestibility in vitro of roughages. Anini Feed Sci Technol 7:61–70

Reddy DV (1996) Evaluation of rice straw-poultry droppings-based rations supplemented with graded levels of rice bran in fistulated buffaloes. Anim Feed Sci Technol 58:227–237

Rodrigues MAM, Pinto P, Bezerra RMF, Dias AA, Guedes CVM, Cardoso VMG, Cone JW, Ferreira LMM, Colaco J, Sequeira CA (2008) Effect of enzyme extracts isolated from white rot fungi on chemical composition and in vitro digestibility of wheat straw. Anim Feed Sci Technol 141:326–338

Saadulah M, Haque M, Dolberg F (1981) Treatment of rice straw with lime. Trop Anim Prod 6:116–120

Sahnounea S, Besle J, Chenost M, Jouany J, Combes D (1991) Treatment of straw with urea. 1. Ureolysis in a low water medium. Anim Feed Sci Technol:75–93

Sarnklong C, Cone JW, Pellikaan W, Hendriks WH (2010) Utilization of rice straw and different treatments to improve its feed value for ruminants: a review Asian-Aust. J Anim Sci 23(5):680–692

Schiere JB, Ibrahim MNM (1989) Feeding of urea ammonia treated rice straw: a compilation of miscellaneous reports produced by the straw utilization project (Sri Lanka). Pudoc, Wageningen

Selim ASM, Pan J, Takano T, Suzuki T, Koike S (2004) Effect of ammonia treatment on physical strength of rice straw, distribution of straw particles and particle-associated bacteria in sheep rumen. Anim Feed Sci Technol 115:117–128

Shen HS, Ni DB, Sundstøl F (1998) Studies on untreated and urea-treated rice straw from three cultivation seasons: 1. Physical and chemical measurements in straw and straw fractions. Anim Feed Sci Technol 73:243–261

Sirohi SK, Rai SN (1995) Associative effect of lime plus urea treatment of paddy straw on chemical composition and in vitro digestibility. Indian J Anim Sci 65:134–135

Stensig T, Weisbjerg MR, Madsen J, Hvelplund T (1994) Estimation of voluntary feed intake from *in sacco* degradation and rate of passage of DM or NDF. Livestock Prod Sci 39:49–52

Sundstøl F, Coxworth EM (1984) Ammonia treatment. In: Sundstøl F, Owen E (eds) straw and other fibrous byproducts as feed, Development in Animal Veterinary Sciences 14. Elsevier, Amsterdam, pp 196–247

Theander O, Aman P (1984) Anatomical and chemical characteristics. In: Sundstøl F, Owen E (eds) Straw and other fibrous byproducts as feed, Developments in Animal Veterinary Sciences, 14. Elsevier, Amsterdam, pp 45–78

Trach NX, Mo M, Da CX (2001) Effects of treatment of rice straw with lime an/or urea on responses of growing cattle. Livest Res Rural Dev 13(5):2001

Uden P (1988) The effect of grinding and pelleting hay on digestibility, fermentation rate, digesta passage and rumen and faecal particle size in cows. Anim Feed Sci Technol 19:145–157

Vadiveloo J (2000) Nutritional properties of the leaf and stem of rice straw. Anim Feed Sci Technol 83:57–65

Vadiveloo J (2003) The effect of agronomic improvement and urea treatment on the nutritional value of Malaysian rice straw varieties. Anim Feed Sci Technol 108:33–146

Vu DD, Cuong LX, Dung CA, Hai PH (1999) Use of urea-molasses-multi-nutrient block and urea-treated rice straw for improving dairy cattle productivity in Vietnam. Prev Vet Med 38:187–193

Walker HA (1984) Physical treatment. In straw and other fibrous byproducts as feed. In: Sundstøl F, Owen E (eds) Developments in animal veterinary sciences, vol 14. Elsevier, Amsterdam, p 79

Wanapat M, Sommart K, Saardrak K (1996) Cottonseed meal supplementation of dairy cattle fed rice straw. Livest Res Rural Dev 8(3)

Wanapat M, Petlum A, Pimpa O (1999) Strategic supplementation with a high quality feed block on roughage intake, milk yield and composition and economic return in lactating dairy cows. Asian-Aust J Anim Sci 12:901–903

Wang Y, Spratling BM, ZoBell DR, Wiedmeier RD, McAllister TA (2004) Effect of alkali pretreatment of wheat straw on the efficacy of exogenous fibrolytic enzymes. J Anim Sci 82:198–208

Warly L, Matsui T, Harumoto T, Fujihara T (1992) Study on the utilization of rice straw by sheep: part I. The effect of soybean meal supplementation on the eating and rumination behavior. Asian-Aust J Anim Sci 5:695–698

Yamakava M, Okamnto HA (1992) Effect of incubation with edible mushroom, *Pleurotus ostreatus*, on voluntary intake and digestibility or rice bran by sheep. Anim Feed Sci Technol 63:133–138

Yang WZ, Beauchemin KA, Rode LM (2000) A comparison of methods of adding fibrolytic enzymes to lactating cow diets. J Dairy Sci 83:2512–2520

Zadrazil F (1977) The conversion of straw into feed by basidiomycetes. Euro J Appl Microbiol Bioteclinol 4:291–294

Zayed MS (2018) Enhancement the feeding value of rice straw as animal fodder through microbial inoculants and physical treatments. Intl J Recycl Organic Waste Agric 7:117–124. https://doi.org/10.1007/s40093-018-0197-7

Zhu S, Wu Y, Yu Z, Liao J, Zhang Y (2005) Pretreatment by microwave/alkali of rice straw and its enzymic hydrolysis. Process Biochem 40:3082–3086

Chapter 8
Rice Straw Incorporation Influences Nutrient Cycling and Soil Organic Matter

Pauline Chivenge, Francis Rubianes, Duong Van Chin, Tran Van Thach, Vu Tien Khang, Ryan R. Romasanta, Nguyen Van Hung, and Mai Van Trinh

Abstract Rice straw incorporation is labor-intensive and influences greenhouse gas emissions but can increase soil organic carbon (C) and recycle nutrients. Rice straw contains about 80, 40, and 30% of the potassium (K), nitrogen (N), and phosphorus (P), respectively, taken up by rice and thus its incorporation can reduce the fertilizer requirement of the subsequent crop. However, because of rice straw's low quality, its decomposition is slow. So, the timing of this operation, in combination with water management, becomes important. Composting rice straw with the addition of farmyard manure can improve quality and nutrient supply. Similarly, biochar from thermal combustion of rice straw for energy production can be added to the soil to improve soil organic C. This chapter highlights the benefits derived from incorporating straw into the soil. Alternative forms of straw that can be used by farmers, depending on local situations, are discussed.

Keywords Soil organic carbon · Straw management · Nutrient supply · Micronutrients · Decomposition

---

P. Chivenge (✉) · F. Rubianes · R. R. Romasanta · N. V. Hung
International Rice Research Institute (IRRI), Los Baños, Laguna, Philippines
e-mail: p.chivenge@irri.org; F.Rubianes@irri.org; r.romasanta@irri.org; hung.nguyen@irri.org

D. Van Chin
Dinh Thanh Agricultural Research Center, Loc Troi Group, Hoa Tan Hamlet, Thoai Son District, An Giang, Vietnam

T. Van Thach
Dinh Thanh Agricultural Research Center (DTARC), Loc Troi Group, Hoa Tan Hamlet, Thoai Son District, An Giang, Vietnam

V. T. Khang
Cuu Long Rice Research Institute, Can Tho City, Vietnam

M. Van Trinh
Institute of Agricultural Environment, Nam Tu Liem District, Hanoi, Vietnam

M. Gummert et al. (eds.), *Sustainable Rice Straw Management*,
https://doi.org/10.1007/978-3-030-32373-8_8

8.1 Introduction

Rice straw management after harvest is an important component of the rice production cycle, particularly in Asia where 90% of the world's rice is produced, and, consequently, where the bulk of the straw is produced. Traditionally in Asia, rice is manually harvested by cutting and carrying to a central threshing location for separation of grain and straw, with only a small portion of straw retained in the field. Straw was considered a waste product and was either burned or used for other purposes such as fodder or animal bedding. However, the increasing use of combine harvesters in the region has resulted in large amounts of rice straw being left in field. The in-situ incorporation of rice straw in the soil has been shown to contribute to recycling of nutrients and increasing soil organic carbon (C) and yields of subsequent crops (Bijay-Singh et al. 2004; Gupta et al. 2007).

While rice straw contains significant amounts of nutrients, its incorporation into the soil is labor-intensive and affects seedbed preparation and crop establishment. This makes land preparation expensive compared to the common practice of open-field burning. On the other hand, straw burning releases particulate matter into the atmosphere, which is associated with air pollution and human respiratory ailments. This has led to bans on open-field straw burning in most major rice-producing countries, although such policies have been largely difficult to enforce. The production of two or three rice crops annually results in the production of large quantities of straw, with little turnaround time between crops, particularly where three crops are grown annually. This results in limited decomposition of the straw when incorporated, with potential negative effects on nutrient availability and use efficiency of applied fertilizers for the subsequent crop (Bijay-Singh et al. 2004; Dobermann and Fairhurst 2000). Depending on the type of water management following straw incorporation, greenhouse gas emissions (GHEs) can also increase (Sander et al. 2014).

Large amounts of rice straw left in the field have posed challenges in rice-growing areas because of the need for mechanization and multiple tillage operations to enable effective incorporation of the straw into the soil. Similarly, the adoption of no-till in rice-cropping systems has been limited by the presence of the large amounts of straw on the soil surface where combine harvesters are used. However, equipment innovations have been developed, such as the Happy Seeder, to enable direct seed drilling while cutting the standing stubble into mulch (Sidhu et al. 2007). Straw incorporation, nonetheless, benefits the next crop and ecosystem services, in general, depending on management practices and the cropping system employed. Incorporating straw in rice fields serves as a source of food for an array of fauna that use rice fields as a habitat. For example, Schmidt et al. (2015) noted that rice straw provides substrate to promote biodiversity through flourishing of invertebrates that decompose the straw, which in turn enhances nutrient cycling in paddy soils. This chapter provides information on the benefits and challenges associated with incorporating rice straw into the soil, including alternative forms through which rice straw can be used as a soil amendment.

8.2 Different Components of Rice Straw

At harvest, rice biomass includes rice straw and paddy grain, which is partitioned into milled grain, bran, and husks after milling. On average, particularly for modern high-yielding varieties, the harvest index (i.e., grain dry weight: total plant biomass dry weight) of rice is between 0.45 and 0.50 (Dobermann and Fairhurst 2000). Thus, roughly for every ton of rice grain produced, a ton of rice straw is also produced. Rice straw usually has higher moisture content than the husks and bran because it is obtained during harvest whereas the bran and husks are drier because they are obtained after dried paddy is milled. When rice is harvested, the straw is removed from the field for threshing to separate grains while the stalk and the stubble remain in the field. The amount of stubble depends on the height at which the straw is cut. The stubble is usually retained in situ and is either burned or incorporated into the soil during land preparation. Rice husks are the parts that are removed when milling paddy to make brown rice while the bran is the layer that is then removed when brown rice is polished to make white rice. Depending on the processing, the different rice plant components other than the grain can be collected and used for composting, biofuel production, or other purposes. Composted materials can be returned to the same soil or applied to the fields of other crops, usually of high value, resulting in the export of nutrients.

8.3 Forms in Which Rice Straw Is Returned to the Soil

Due to ease of management, the main form in which rice straw is currently returned to the soil is by incorporating burned ash after open-field burning of stubble following harvest. However, with the banning of straw burning in most rice-growing countries, incorporating fresh rice residue serves as the alternative for returning rice straw to rice fields. While this is associated with benefits of improving soil quality and recycling nutrients, it can have negative impacts on the environment through release of GHGs, depending on management. Alternatively, the straw is collected from the field and mixed with other inputs such as livestock manure, green manure, or household waste to make compost, which is more decomposed and of higher quality compared to fresh straw. In other instances, earthworms and microorganisms are added to the straw to enhance decomposition to make vermicompost (see Chap. 3). Compost and vermicompost can be made from fresh or spent straw from mushroom production, but farmers generally prefer to use them for high-value crops such as vegetables rather than for rice production. The straw can be used in energy production and the carbonized rice straw that remains can be used as a soil amendment (Haefele et al. 2011).

8.4 Straw Effects on Soil Properties

8.4.1 Nutrient Cycling

The incorporation of rice straw can improve soil quality through enhanced nutrient cycling and soil organic C sequestration. Straw incorporation has been shown to enhance nutrient recycling and provide soil fertility benefits (Dobermann and Fairhurst 2000; Ponnamperuma 1984; Yadvinder-Singh et al. 2004). The straw is the main organic material that is available for most rice farmers and serves as an important source of K. Ponnamperuma (1984) indicated that, at harvest, the straw contains 0.57% N, 0.07% P_2O_5, 1.5% K_2O, 0.1% (sulfur) S, and 5% silicon (Si). In agreement, Dobermann and Fairhurst (2002) also showed that rice straw at harvest can contain 0.5–0.8% N, 0.07–0.12% P_2O_5, 1.16–1.66% K_2O, 0.05–0.1% S, and 4–7% Si. This translates to about 40%, 30–35%, 80–85%, 40–50%, and 80%, respectively, of the N, P_2O_5, K_2O, S, and Si taken up by the plant. The addition of rice straw was shown to improve soil pH, soil organic C, and nutrient content compared to the initial conditions in a study conducted in Vietnam (Table 8.1) (Thanh et al. 2016). In that study, the incorporation of rice straw resulted in greater increases in soil organic C, pH, and nutrient contents compared to addition of ash from burned straw, although the increase in N was small. Due to the low N content in straw, large quantities would be needed to supply adequate amounts of N. However, the straw has to decompose before the nutrients can become available for uptake and the rate of decomposition and supply of nutrients depends on soil type and season. Additionally, only a proportion of the nutrients become available in the season of application. For example, in a study on an alluvial soil in Vietnam, about 67 to 69% of the rice straw had decomposed by the time the plant had reached physiological maturity (Thuan and Long 2010).

The availability of nutrients is affected by the low quality of rice straw, with a high C:N ratio, resulting in slow decomposition and mineralization of nutrients, particularly short-term availability of N and to some extent P (Thuy et al. 2008). The C:N ratio of an organic material determines its quality, with high C:N ratio representing low quality and a slow rate of decomposition, whereas low C:N ratio represents high quality with a faster decomposition. The addition of rice straw to wet soil results in temporary immobilization of nitrogen, making it unavailable and affecting rice yields (Bird et al. 2001). Apart from low straw quality, N availability following

Table 8.1 Effects of rice straw and ash from rice straw on soil chemical properties

Treatment	pH_{KCl}	SOC[a]	N	P_2O_5	K_2O
		(%)			
Before experiment	4.10	0.80	0.08	0.034	0.52
Ash from 5 t ha^{-1} rice straw	4.32	1.09	0.09	0.046	0.60
5 t ha^{-1} fresh rice straw	4.40	1.19	0.11	0.041	0.55

Source: Thanh et al. (2016)
[a]*SOC* soil organic C

incorporation of the straw is affected by the accumulation of phenolic compounds that are formed under the straw's anaerobic decomposition (Olk et al. 2006). These phenolic compounds tend to bind the N in the soil making it unavailable for plant uptake.

Nonetheless, the long-term incorporation of crop residues in flooded rice soil can increase soil organic matter, total N, and soil biological activity (Yadvinder-Singh et al. 2004). Continuous incorporation of crop residues after each crop can eventually increase the N-supplying capacity of rice soils (Eagle et al. 2000). In a study in Vietnam, soil N increased from 0.65% to 0.085% following 9 years of cropping with incorporation of rice straw while straw removal caused a decline in soil N (Thuan and Long 2010). The benefits of incorporated residues on soil organic matter and soil N supply, however, seldom translate into increased yield or profit for flooded rice (Bijay-Singh et al. 2008). However, Thanh et al. (2016) observed that N fertilizer requirement was reduced by about 20% in a long-term study with rice straw incorporation.

Timings of straw incorporation and water management are important considerations for effective use of straw as a nutrient resource (Dobermann and Fairhurst 2002; Witt et al. 2000). Rice straw should be incorporated in dry soil at least 3 weeks before sowing or transplanting the next crop to allow the straw to decompose aerobically. This minimizes the negative effects of anaerobic decomposition, which results in release of phenolic compounds (Olk et al. 2006), and methane emissions (Sander et al. 2014), while allowing for decomposition and mineralization of the nutrients, making them available during plant growth. In a cropping system comparing rice–rice to rice–maize, N supply was greater when residue incorporation took place 63 rather than 14 days before planting wet season rice; this was associated with greater rice yields (Witt et al. 2000). This points to the need for time for straw decomposition before the nutrients become available for plant uptake. Rice straw compost, on the other hand, because it is more decomposed before its application in the soil, has greater nutrient availability compared to raw rice straw. However, farmers prioritize the use of compost on higher-value crops such as vegetables than on rice.

Rice straw can serve as an important source of S, which is particularly important in situations where S-free fertilizers are used (Dobermann and Fairhurst 2002). It also serves as an important source of micronutrients including zinc (Zn), but its long-term application can decrease the availability of Zn (Yadvinder-Singh et al. 2005). Rice straw is also important for P recycling. For example, in a 4-year study in India, P balances were negative where rice and wheat straw were removed or burned (Gupta et al. 2007). The mineral P dynamics were improved where P fertilizers were added. Similarly, Gangwar et al. (2006) observed greater concentrations of plant available P when rice straw was incorporated compared to when it was removed in a 3-year rice–wheat study. Continuous addition of biochar made from rice straw on a degraded soil in Soc Son District of Hanoi resulted in an increase in soil pH, cation exchange capacity, and soil organic C after four seasons (Table 8.2A) (Trinh et al. 2011). Biochar has high soil pH and tends to have a liming effect in soil, while it is also stable and decomposes slowly, resulting in an increase in soil organic

Table 8.2 Soil properties before and after four rice seasons of continuous application of (A) biochar from rice husks or straw compared to farmyard manure and mineral NPK fertilizer in Soc Son District, Hanoi, Vietnam, and (B) NPK fertilizer compared to biochar, compost from rice straw, and compost from rice straw mushroom solid waste on three sites; Thai Binh, Hung Yen, and Hai Duong provinces in Red River Delta, Vietnam

Treatment	pH_{H20}	N (%)	P_2O_5	K_2O	SOC[a]	CEC (cmolc$_+$ kg^{-1})	Ca	Mg
A. Trinh et al. (2011)								
Before experiment	5.02	0.13	0.07	0.22	1.33	9.24	2.04	0.21
No fertilizer (control)	5.20	0.10	0.09	0.21	1.10	10.4	2.74	0.09
NPK fertilizer	5.25	0.15	0.11	0.27	1.27	10.6	3.51	0.25
Farmyard manure (10 t ha^{-1})	5.24	0.15	0.10	0.28	1.62	11.5	3.59	0.35
Rice husk biochar (1.5 t ha^{-1})[b]	5.30	0.17	0.11	0.33	1.83	11.6	3.48	0.35
Rice husk biochar (3.0 t ha^{-1})	5.31	0.17	0.11	0.31	1.89	14.7	3.82	0.48
Rice straw biochar (1.5 t ha^{-1})[b]	5.39	0.17	0.13	0.32	1.85	13.6	3.55	0.36
Rice straw biochar (3.0 t ha^{-1})	5.45	0.17	0.12	0.32	1.79	14.5	4.02	0.45
Rice straw biochar (4.5 t ha^{-1})	5.40	0.18	0.14	0.35	1.82	15.2	4.11	0.48
B. Trinh et al. (2014)								
NPK fertilizer	5.54	0.21	0.18	1.56	2.05	14.29		
Biochar	5.82	0.23	0.20	1.49	2.31	16.52		
Compost from straw	5.86	0.22	0.22	1.57	2.35	16.24		
Compost from straw mushroom waste	5.81	0.21	0.20	1.72	2.18	17.29		

In all the treatments for both studies, NPK fertilizers were added, except for the no fertilizer control treatment

[a]*SOC* soil organic C

[b]The amount of C in this treatment is equal to amount of C in 10 tons of Farmyard Manure

C. Soil water and nutrient holding capacity were also increased, likely due to the porous nature of biochar, with an indirect increase in rice yield. A separate study conducted in Thai Binh, Hung Yen, and Hai Duong provinces in the Red Delta in Vietnam showed significant improvement in soil fertility following the addition of rice straw and products from rice straw (biochar, compost, compost from rice straw mushroom; Table 8.2B) (Trinh et al. 2014).

8.4.2 Soil Organic Carbon

The sequestration of organic C in soil is generally considered a win-win situation because of its contribution to the mitigation of GHGs from the atmosphere while improving soil quality. Burning of straw, on the other hand, results in combustion of the C in the straw and loss into the atmosphere, associated with production of GHGs (see Chap. 10). Soil organic C is an important component of the global C cycle and is considered an indicator of soil quality and a measure of sustainability. Soil organic C is the main component of soil organic matter, which plays an important role in the supply of nutrients and improves biological and physical properties of the soil. Rice straw incorporation has been shown to increase soil organic C (Tables 8.1 and 8.2). However, the role of soil organic C in rice soils remains debatable. While soil organic C is considered important on one hand, the increased GHG emissions associated with increased soil organic C can contribute to climate change. Thus, there is need to evaluate the tradeoffs and synergies of soil organic C sequestration in rice soils.

Soil organic C has been shown to be stable under intensive rice cropping, even when straw is removed from the field. Soil organic C was shown not to change in a 50-year, long-term continuous cropping experiment at the International Rice Research (IRRI) in the Philippines where three rice crops were grown annually with the removal of all aboveground biomass even without the addition of N fertilizer (Pampolino et al. 2008). This is in contrast to systems where rice is rotated with an upland crop, e.g., Majumder et al. (2008) observed a decline in soil organic C when no residues were added in a rice–wheat cropping system in India. In a 9-year study in Bac Giang Province in Vietnam, soil organic C did not change with straw removal, but the addition of straw increased soil organic C from 1.28% to 1.65% (Thuan and Long 2010). Alberto et al. (2015) showed a cumulative effect of continuous straw incorporation in a lowland rice soil, likely due to slower organic matter decomposition. However, the addition of straw increases soil organic C (Bi et al. 2009; Yadvinder-Singh et al. 2005), particularly in rainfed upland rice systems (Naklang et al. 1999) or where lowland rice is rotated with an upland crop. In a rice–wheat system, Gangwar et al. (2006) observed greater soil organic C and infiltration when 5 t ha^{-1} rice straw was incorporated in the soil than when it was removed or burned.

8.5 Rice Straw Effects on Yield

While yield increases are expected with the retention of crop residues in upland cropping systems, in lowland rice the benefits when compared to straw removal are small particularly in the short-term. Under continuous flooded rice, the retention of rice straw has not been shown to increase rice yield. This might be due to the low-quality nature of rice straw with a high C:N ratio, which results in N immobilization and hence poor availability for plant uptake. Additionally, anaerobic decomposition of organic materials has been shown to trigger production of phenolic compounds that also renders N to be unavailable and affect crop growth. Incorporation of rice straw on three different soil types did not increase rice yield and this was attributed to an increase in toxic substances and organic acids (Hoi et al. 2009). This is particularly important when the straw has not been given adequate time for decomposition. However, long-term benefits of straw incorporation on rice yield can be significant. A summary of some studies conducted in the Philippines and Vietnam shows yield benefits from straw incorporation (Table 8.3).

In a long-term study in the Mekong Delta in Vietnam (Watanabe et al. 2009) observed that the application of 6 Mg ha^{-1} rice straw compost (fresh weight)

Table 8.3 A synthesis of straw management effects on rice yield in some studies in the Philippines and Vietnam

Treatment	Rate Kg ha^{-1}	Location/ country	Seasons[a]	Soil description	Grain yield[b] Mg ha^{-1}	Source
No straw		Philippines		Maahas clay	3.2 b	Banta and Mendoza (1984)
Straw burned					3.4 b	
Straw incorporated					4.1 a	
Straw composted					4.2 a	
No straw		Philippines			77	Banta and Mendoza (1984)
Straw incorporated	5000				112	
No straw		IRRI, Philippines	10 DS	Silty clay	3.79 e	Cassman et al. (1996)
Straw incorporated	116			Silty clay	4.70 d	
No straw		Victoria, Philippines	9 DS	Clay	3.83 d	
Straw incorporated	116			Clay	4.93 c	
No straw		IRRI, Philippines	11 WS	Silty clay	3.39 d	
Straw incorporated	58			Silty clay	3.57 c	
No straw		Victoria, Philippines	9 WS	Clay	3.22 d	
Straw incorporated	58			Clay	3.56 c	
No straw		Can Tho – Vietnam	3 DS	Fluvaquentic Humaquepts	2.94	Tuyen and Tan (2001)
Straw burned					2.96	
Straw incorporated					2.72	
No straw		Can Tho – Vietnam	3 DS		5.70	
Straw burned					5.50	
Straw incorporated					5.59	
Raw rice straw		An Giang Province, Vietnam	1 DS		2.83 c	Son et al. (2013)
Composted rice straw					2.95 c	
Raw rice straw +70% NPK					4.71 b	
Composted rice straw +70% NPK					5.32 a	
Burned rice straw +70% NPK					4.77 b	
Burned rice straw +100% NPK					5.11 ab	
Raw rice straw +100% NPK					5.30 a	
Composted rice straw+100% NPK					5.33 a	

(continued)

Table 8.3 (continued)

Treatment	Rate Kg ha^{-1}	Location/ country	Seasons[a]	Soil description	Grain yield[b] Mg ha^{-1}	Source
No straw; no additional NPK		Can Tho – Vietnam	4	Typic Humaquept	3.79[c]	Watanabe et al. (2013)
Straw incorporated; no additional NPK	6000				4.81[c]	
Straw incorporated + NPK (40,12,12 kg ha^{-1})	6000				6.06[c]	
Straw incorporated + NPK (60,18,18 kg ha^{-1})	6000				6.04[c]	
No straw; no NPK		Can Tho – Vietnam	4		4.19[d]	Watanabe et al. (2017)
Straw incorporated; no NPK	6000				4.91[d]	
No straw + inorganic 40% NPK					5.33[d]	
Straw incorporated +40% NPK	6000				5.95[d]	
No straw +60% inorganic NPK					5.48[d]	
Straw incorporated +60% NPK	6000				5.90[d]	
No straw +100% NPK					4.94[d]	

[a]Seasons means number of seasons; *DS* dry season; *WS* wet season; numbers without letters means the season type is not specified
[b]Treatment means followed by the same lowercase letter within the same study are not significantly different at $p < 0.05$
[c]Average grain yield over four seasons from 2011 to 2013
[d]Average grain yield over four seasons from 2009 to 2011

increased rice yield where no mineral fertilizer was applied in the wet season (Table 8.3). In the same study, they also observed positive effects of rice straw compost on physical soil properties including a lower penetration resistance compared to where no compost was applied. In China, rice yield was greater with rice straw incorporation than removal under conventional tillage where no nitrogen fertilizer was added (Xu et al. 2010). A 3-year study conducted across three rice-growing sites in Asia showed little or no benefit of incorporated rice or wheat straw for the succeeding crop (Thuy et al. 2008). However, at a site in India the incorporation of rice straw 20 days before sowing wheat without N fertilization significantly decreased wheat yield but increased yield of rice that followed after wheat. In contrast, in East China incorporation of rice straw increased wheat yield by about 28% compared to no straw control, but had no significant effects on rice yield (Zhang et al. 2015).

A study conducted on a rice–wheat cropping system in India over 4 years showed greater yields of wheat where rice and wheat straw were incorporated compared to where it was removed or burned (Gupta et al. 2007). In contrast, straw management did not affect rice yields. A meta-analysis conducted in China showed that retention of rice straw increased rice yields by 5.2%, but that the yield benefit increased with duration of straw incorporation, i.e., time after repeated application (Huang et al. 2013). In a study in India, wheat grain yield was 13% greater when rice straw was incorporated in the soil compared to when it was removed. A recent study in India showed that cereal yields, i.e., rice, wheat, and maize, are greater where straw is incorporated than where it is removed under a diversified cropping system and different combinations of tillage and crop establishment methods (Nandan et al. 2018). While the rice yield increased from 3.0% to 8.2%, the yield benefits were greater for maize and wheat, likely due to a combination of moisture conservation under upland conditions and nutrient contribution from the incorporated straw.

8.6 Paddy Soil Degradation Associated with Straw Removal

While straw incorporation is labor-intensive and requires machinery for effective mixing with the soil, especially where rice yields are high, straw burning has negative impacts on soil fertility and soil organic C (Prasad et al. 1999; Surekha et al. 2003). The heat and duration of fire; soil moisture, both at the time of burning and during tillage; the time elapsed and climatic condition between burning and tillage; and the chemical, physical, and biological properties of the soil will all influence the change in soil properties resulting from burning rice straw. The impact of straw burning on soil fertility accumulates over time. Ponnamperuma (1984) highlighted the need to consider experimenting with duration when drawing conclusions about the sustainability of straw burning. However, research indicates that the advantages of burning are offset by the disadvantages, including nutrient loss, depletion of soil organic C, and reduction in the presence of beneficial soil biota (Mandal et al. 2004). Nonetheless, the ash from burning rice straw is rich in K and, thus, the practice of straw burning resulted in recycling of K, but with a loss of N and P.

Potassium recycling is influenced by straw management since more than 80% of the K taken up by rice is in straw (Dobermann and Fairhurst 2000). Consequently, K deficiencies are common in soils where rice straw is removed. In a study conducted under rainfed conditions in Thailand, Whitbread et al. (2003) calculated negative K and S balances when rice straw was removed. Similarly, straw management influences Si, which has been shown to be beneficial in rice growth. Si dynamics in the soil is affected by straw management (Seyfferth et al. 2013; Wickramasinghe and Rowell 2006). For example, soil-available Si was low in Vietnam where crop residues are removed from the fields while in the Philippines there is high availability due to crop residue retention (Settele et al. 2018). Removal of rice straw from the field has been practiced widely in South Asia and has been associated with K and Si deficiencies influencing rice productivity (Wickramasinghe and Rowell 2006). Removal of rice straw from the field can cause numerous direct and indirect adverse impacts on the ecosystem including depletion of soil organic C. Important direct impacts of removal are low input of C biomass, reduction in nutrients/elemental cycling and decrease in food/energy source for soil biota along with the attendant decline in soil quality (Vijayaprabhaka et al. 2017).

8.7 Constraints, Trends, and Recommendations

Increasingly, rice straw is incorporated in fields where rice is grown. This is more so with the increasing use of combine harvesters that leave all the crop residues in the field compared to the traditional manual harvesting where only a portion of the straw was retained in the field. There are benefits associated with the incorporation of rice straw in the soil, including greater yields, nutrient cycling, soil organic matter build-up, and a general benefit on ecosystem services. However, there is a need to consider timing of application and water management to maximize the benefits of straw in the soil and reduce the negative effects such as production of GHG emissions or release of phenolic compounds that affect nutrient availability. This brings into question the practicality of straw incorporation on intensive systems, e.g., triple rice cropping where there is little time between crops to allow for aerobic decomposition. This suggests that solutions for straw management need to be tailored to suit farmer conditions. In situations where triple rice cropping is mechanized, there is a need to consider cost-effective alternatives to straw management with documentation of the experiences to enable extension workers to give informed recommendations to farmers.

In cases where combine harvesters are used, leaving large quantities of rice residue in the field, the practice of zero tillage with the use of seed drills is affected by residues clogging the machinery. Alternative equipment should be considered. There is also a need for studies to determine proportions of straw that can be left on the field with minimum effects on land preparation and establishment of the next crop. Options for collection of straw from the field, especially where combine harvesters are used, have been discussed elsewhere in this book, but decisions on whether to collect or leave straw in the field need to take into consideration the trade-offs of the different options. Nonetheless, there are benefits, in general, of returning rice straw to the field for sustainable cycling of nutrients and improving crop yields.

References

Alberto MCR, Wassmann R, Gummert M, Buresh RJ, Quilty JR, Correa TQ Jr, Centeno CAR, Oca GM (2015) Straw incorporated after mechanized harvesting of irrigated rice affects net emissions of CH_4 and CO_2 based on eddy covariance measurements. Field Crop Res 184:162–175

Banta S, Mendoza CV (eds) (1984) Organic matter and rice. International Rice Research Institute http://books.irri.org/9711041049_content.pdf

Bi L, Zhang B, Liu G, Li Z, Liu Y, Ye C, Yu X, Lai T, Zhang J, Yin J (2009) Long-term effects of organic amendments on the rice yields for double rice cropping systems in subtropical China. Agric Ecosyst Environ 129:534–541

Bijay-Singh YS, Ladha JK, Khind CS, Khera TS, Bueno CS (2004) Effects of residue decomposition on productivity and soil fertility in rice–wheat rotation. Soil Sci Soc Am J 68:854–864

Bijay-Singh YS, Shan Y, Johnson-Beebout S, Yadvinder-Singh BR (2008) Crop residue management for lowland rice-based cropping systems in Asia. Adv Agron 98:117–199

Bird JA, Horwath WR, Eagle AJ, Van Kessel C (2001) Immobilization of fertilizer nitrogen in rice. Soil Sci Soc Am J 65:1143–1152

Cassman K, De Datta S, Amarante S, Liboon S, Samson M, Dizon M (1996) Long-term comparison of the agronomic efficiency and residual benefits of organic and inorganic nitrogen sources for tropical lowland rice. Exp Agric 32:427–444

Dobermann A, Fairhurst TH (2000) Rice: nutrient disorders and nutrient management. International Rice Research Institute http://books.irri.org/9810427425_content.pdf

Dobermann A, Fairhurst T (2002) Rice straw management. Better Crops Intl 16:7–11

Eagle AJ, Bird JA, Horwath WR, Linquis BA, Brouder SM, Hill JE, van Kessel C (2000) Rice yield and nitrogen utilization efficiency under alternative straw management practices. Agron J 92(6):1096–1103

Gangwar KS, Singh KK, Sharma SK, Tomar OK (2006) Alternative tillage and crop residue management in wheat after rice in sandy loam soils of Indo-Gangetic plains. Soil Tillage Res 88:242–252

Gupta R, Ladha J, Singh J, Singh G, Pathak H (2007) Yield and phosphorus transformations in a rice–wheat system with crop residue and phosphorus management. Soil Sci Soc Am J 71:1500–1507

Haefele S, Konboon Y, Wongboon W, Amarante S, Maarifat A, Pfeiffer E, Knoblauch C (2011) Effects and fate of biochar from rice residues in rice-based systems. Field Crop Res 121:430–440

Hoi NT, Ve NB, Tan PS, Giau TQ (2009) Effects of incorporated rice straw into the soil on rice growth and yield. Journal of Sciences. Can Tho University

Huang S, Zeng Y, Wu J, Shi Q, Pan X (2013) Effect of crop residue retention on rice yield in China: a meta-analysis. Field Crop Res 154:188–194

Majumder B, Mandal B, Bandyopadhyay P, Gangopadhyay A, Mani P, Kundu A, Mazumdar D (2008) Organic amendments influence soil organic carbon pools and rice–wheat productivity. Soil Sci Soc Am J 72:775–785

Mandal KG, Misra AK, Hati KM, Bandyopadhyay KK, Ghosh PK, Mohanty M (2004) Rice residue-management options and effects on soil properties and crop productivity. J Food Agric Environ 2:224–231

Naklang K, Whitbread A, Lefroy R, Blair G, Wonprasaid S, Konboon Y, Suriya-arunroj D (1999) The management of rice straw, fertilizers and leaf litters in rice cropping systems in Northeast Thailand. Plant Soil 209:2–128

Nandan R, Singh S, Kumar V, Singh V, Hazra K, Nath C, Malik R, Poonia S, Solanki CH (2018) Crop establishment with conservation tillage and crop residue retention in rice-based cropping systems of eastern India: yield advantage and economic benefit. Paddy Water Environ 16:477–492

Olk DC, Cassman KG, Schmidt-Rohr K, Anders MM, Mao JD, Deenik JL (2006) Chemical stabilization of soil organic nitrogen by phenolic lignin residues in anaerobic agroecosystems. Soil Biol Biochem 38:3303–3312

Pampolino MF, Laureles EV, Gines HC, Buresh RJ (2008) Soil carbon and nitrogen changes in long-term continuous lowland rice cropping. Soil Sci Soc Am J 72:798–807

Ponnamperuma F (1984) Straw as a source of nutrients for wetland rice. Organic Matter Rice 117:136

Prasad R, Gangaiah B, Aipe K (1999) Effect of crop residue management in a rice–wheat cropping system on growth and yield of crops and on soil fertility. Exp Agric 35:427–435

Sander BO, Samson M, Buresh RJ (2014) Methane and nitrous oxide emissions from flooded rice fields as affected by water and straw management between rice crops. Geoderma 235:355–362

Schmidt A, Auge H, Brandl R, Heong KL, Hotes S, Settele J, Villareal S, Schädler M (2015) Small-scale variability in the contribution of invertebrates to litter decomposition in tropical rice fields. Basic Appl Ecol 16:674–680

Settele J, Heong KL, Kühn I, Klotz S, Spangenberg JH, Arida G, Beaurepaire A, Beck S, Bergmeier E, Burkhard B (2018) Rice ecosystem services in Southeast Asia. Springer

Seyfferth AL, Kocar BD, Lee JA, Fendorf S (2013) Seasonal dynamics of dissolved silicon in a rice cropping system after straw incorporation. Geochim Cosmochim Acta 123:120–133

Sidhu H, Humphreys E, Dhillon S, Blackwell J, Bector V (2007) The Happy Seeder enables direct drilling of wheat into rice stubble. Aust J Exp Agric 47:844–854

Son TTN, Thu TTA, Nam NN, Man LH (2013) Influence of rice straw treated by indigenous *Trichoderma* spp. on soil fertility, rice grain yield, and economic efficiency in the Mekong Delta. Methods 60:145–152

Surekha K, Kumari AP, Reddy MN, Satyanarayana K, Cruz PS (2003) Crop residue management to sustain soil fertility and irrigated rice yields. Nutr Cycl Agroecosyst 67:145–154

Thanh ND, Hoa HTT, Hung HC, Nhi PTP, Thuc DD (2016) Effect of fertilizer on rice yield improvement in coastal sandy soil of Thua Thien Hue province. Hue Univ J Sci Agric Rural Dev 119

Thuan HN, Long DT (2010) Use rice straw from previous season for the following season on degraded soil, Bac Giang province, Vietnam. J Sci Dev, Hanoi Agric University 8:843–849

Thuy NH, Shan Y, Bijay-Singh WK, Cai Z, Yadvinder-Singh BRJ (2008) Nitrogen supply in rice-based cropping systems as affected by crop residue management. Soil Sci Soc Am J 72:514–523

Trinh MV, Cuong V, Quynh VD, Thu NTH (2011) Study to produce biochar from rice straw and rice husk to enhance soil fertility, crop yield, and mitigate GHG emission. J Vietnam Agric Sci Technol 3

Trinh MV, Loan BTP, Van The T (2014) Developing crop residue collection, treatment of crop residue to reduce green house gas emission in rural area in Red River Delta, Report results of project from Ministry of Agriculture and Rural Development period 2011–2014

Tuyen TQ, Tan PS (2001) Effects of straw management and tillage practices on soil fertility and grain yield of rice. Omonrice 9:74–78

Vijayaprabhaka A, Nalliah Durairaj S, Vinoth Raj J (2017) Effect of rice straw management options on soil available macro and micro nutrients in succeeding rice field. Intl J Chem Stud 5:410–413

Watanabe T, Man LH, Vien DM, Khang VT, Ha NN, LinhTB IO (2009) Effects of continuous rice straw compost application on rice yield and soil properties in the Mekong Delta. Soil Sci Plant Nutr 55:754–763

Watanabe T, Luu HM, Nguyen NH, Ito O, Inubushi K (2013) Combined effects of the continual application of composted rice straw and chemical fertilizer on rice yield under a double rice cropping system in the Mekong Delta, Vietnam. Japan Agric Res Quart JARQ 47:397–404

Watanabe T, Luu HM, Inubushi K (2017) Effects of the continuous application of rice straw compost and chemical fertilizer on soil carbon and available silicon under a double rice cropping system in the Mekong Delta, Vietnam. Japan Agric Res Quart JARQ 51:233–239

Whitbread A, Blair G, Konboon Y, Lefroy R, Naklang K (2003) Managing crop residues, fertilizers and leaf litters to improve soil C, nutrient balances, and the grain yield of rice and wheat cropping systems in Thailand and Australia. Agric Ecosyst Environ 100:251–263

Wickramasinghe DB, Rowell DL (2006) The release of silicon from amorphous silica and rice straw in Sri Lankan soils. Biol Fertil Soils 42:231–240

Witt C, Cassman K, Olk D, Biker U, Liboon S, Samson M, Ottow J (2000) Crop rotation and residue management effects on carbon sequestration, nitrogen cycling, and productivity of irrigated rice systems. Plant Soil 225:263–278

Xu Y, Nie L, Buresh RJ, Huang J, Cui K, Xu B, Gong W, Peng S (2010) Agronomic performance of late-season rice under different tillage, straw, and nitrogen management. Field Crop Res 115:79–84

Yadvinder-Singh, Bijay-Singh, Timsina J (2005) Crop residue management for nutrient cycling and improving soil productivity in rice-based cropping systems in the tropics. Adv Agron 85:269–407

Yadvinder-Singh B-S, Ladha J, Khind C, Khera T, Bueno C (2004) Effects of residue decomposition on productivity and soil fertility in rice–wheat rotation. Soil Sci Soc Am J 68:854–864

Zhang L, Zheng J, Chen L, Shen M, Zhang X, Zhang M, Bian X, Zhang J, Zhang W (2015) Integrative effects of soil tillage and straw management on crop yields and greenhouse gas emissions in a rice–wheat cropping system. Eur J Agron 63:47–54

Chapter 9
Rice Straw Management Effects on Greenhouse Gas Emissions and Mitigation Options

Justin Allen, Kristine S. Pascual, Ryan R. Romasanta, Mai Van Trinh, Tran Van Thach, Nguyen Van Hung, Bjoern Ole Sander, and Pauline Chivenge

Abstract Lowland rice is a significant source of anthropogenic greenhouse gas emissions (GHGEs) and the primary source of agricultural emissions for many developing countries in Asia. At the same time, rice soils represent one of the largest global soil organic carbon sinks. Straw management is a key factor in controlling the emissions and mitigation potential of rice primarily by affecting methane (CH_4) from anaerobic decomposition and carbon losses from burning. Achieving climate-smart management of rice while also improving yields and farm profits, however, is challenging due to economic-environmental trade-offs. This balance could be met with appropriate site-specific practices. This chapter discusses these straw management practices that affect yield-scaled GHGEs and mitigation options in different rice environments.

Keywords Greenhouse gas emissions · GHG · Mitigation · Rice straw

---

J. Allen (✉) · R. R. Romasanta · N. V. Hung · P. Chivenge
International Rice Research Institute (IRRI), Los Baños, Laguna, Philippines
e-mail: j.allen@irri.org; r.romasanta@irri.org; hung.nguyen@irri.org; p.chivenge@irri.org

K. S. Pascual
Philippine Rice Research Institute (PhilRice), Muñoz, Nueva Ecija, Philippines

M. Van Trinh
Institute of Agricultural Environment, Nam Tu Liem District, Hanoi, Vietnam

T. Van Thach
Dinh Thanh Agricultural Research Center (DTARC), Loc Troi Group, Hoa Tan Hamlet, Thoai Son District, An Giang, Vietnam

B. O. Sander
International Rice Research Institute (IRRI), Tu Liem District, Hanoi, Vietnam
e-mail: b.sander@irri.org

M. Gummert et al. (eds.), *Sustainable Rice Straw Management*,
https://doi.org/10.1007/978-3-030-32373-8_9

9.1 Introduction

Lowland rice is a major contributor to greenhouse gas emissions (GHGEs) accounting for 10% of global emissions from agriculture (FAO 2015). This number is even higher for Southeast Asia (SEA) where 90% of the world's rice is produced, making up 10–20% of the region's total anthropogenic emissions and 40–60% of its agricultural emissions (UNFCC 2019). Rice is one of the largest sources of anthropogenic CH_4 (GWP[1] = 28) and a major contributor of N_2O (GWP = 265). CO_2 emissions from rice, although large, are considered net-neutral from photosynthesis according to the IPCC 2006 guidelines. CH_4 accounts for around 65% of global CO_2 eq emissions from lowland rice; largely from anaerobic decomposition of straw and crop residue under continuously flooded conditions. The remaining 35% of emissions from rice can be attributed mostly to N_2O from soil N cycling of fertilizer and to a smaller extent N from crop residues (EPA 2013). Rice straw management is, therefore, an important factor in controlling GHGEs from lowland rice-cropping systems.

In addition to emissions, straw management plays an important role in global carbon cycles through soil organic carbon (SOC) sequestration. SOC is an important indicator of soil quality, which suggests its importance in improving farmer adaptation to climate change. It is estimated that rice soils contain the largest SOC stocks among croplands (IPCC 2007; Lal 2004). The potential SOC deposition from returning rice straw to the soil is significant as almost half of the total carbon in rice plant residue is within the straw and stubble (although root C contributes most SOC). The common, yet mostly banned, practice of straw burning reduces the SOC sequestration potential of fresh straw incorporation.

Although returning fresh straw to the field can increase SOC, its sequestration benefits may be outweighed by the increase in CH_4 emissions when applied under flooded conditions due to anaerobic decomposition. Additionally, straw management practices that reduce emissions or improve sequestration are not always advantageous to crop yields. Striking a balance between emissions reduction, carbon sequestration, and crop yields is challenging, but may be achievable with optimal site-specific straw management. The efficiency of this balance can be quantified by yield-scaled emissions and mitigation or NGWP and GHGI,[2] more broadly referred to as climate-smart agriculture (CSA). This chapter discusses in-field/off-field rice straw management options affecting CSA—burning, incorporation, com-

[1]Global warming potential (GWP) is a measure of how much heat a greenhouse gas traps in the atmosphere up to a specific time horizon, relative to carbon dioxide (CO_2).

[2]Net global warming potential (NGWP) can be defined as the radiative properties of all the GHG emissions plus carbon fixation, expressed as CO_2 eq ha^{-1} $year^{-1}$ (Robertson and Grace 2004), while greenhouse gas intensity (GHGI) defines the GWP per unit of crop yield (Mosier et al. 2006)

posting, biochar, and others—under various rice production environments, such as water management, cropping system, and soil type.

9.2 In-Field Straw Management Effects on Emissions and Mitigation

9.2.1 Burning

Open-field burning of rice straw has well-known negative environmental and agronomic impacts due to atmospheric pollution and reduced soil quality. Burning also emits GHGs CO_2, CH_4, and N_2O, along with other trace gases that contribute to tropospheric ozone and the formation of Atmospheric Brown Cloud (ABC)—a cause of severe human health concern (Arai et al. 1998; Gullett and Touati 2003; Lin et al. 2007; Tipayarom and Kim Oanh 2007; Torigoe et al. 2000; Kanokkanjana et al. 2011). Still, studies suggest that the total GHGEs from burning are up to 98% lower than those from fresh straw incorporation in flooded soils due to reductions in CH_4 from straw decomposition (IPCC 2006). This accounting, however, excludes CO_2 emissions, which are considered net neutral from photosynthesis in the IPCC guidelines. When CO_2 is included, the carbon losses from burning reduce the SOC sequestration potential of fresh straw incorporation due to the immediate 90% loss of straw C as CO_2 during combustion (Chen et al. 2019). When this is accounted for, the NGWP from burning is comparable to that of complete fresh straw incorporation (Lu et al. 2010).

SOC sequestration is thus an important component of emissions calculations from burning. For example, a meta-analysis in China compared the effects of burning and straw incorporation on NGWP to include sequestration and found that switching from burning to straw incorporation could mitigate 34.18 Mt. CO_2 eq year^{-1} or 31% of total rice emissions in the country (Lu 2015; Liu et al. 2014). This assumed a large sequestration potential by restoring degraded soils to their maximum SOC storage ability or SOC saturation capacity (EPA 2013). Once saturation was reached, the mitigation potential of straw incorporation diminished. Increasing SOC not only mitigates emissions, but can also substantially improve soil quality, yields, and adaptation to climate change by improving drought tolerance. For example, an only 1% improvement to SOM can double the soil water holding capacity (Fileccia et al. 2014).

Despite the established negative long-term impacts of burning on soil quality, SOC sequestration and air quality, intensive rice farmers prefer burning rice straw due to lower costs, reduced weed and disease carryover, and ease of tillage. Advantages of burning may decline as opportunities increase for off-farm uses and stricter government environmental regulations encourage alternative options.

9.2.2 *Incorporation Rates and Environmental Factors*

9.2.2.1 Water Management

CH_4 emissions from rice are highly dependent on the amount of straw or crop residue returned under continuously flooded conditions (Liu et al. 2014). Because of this, removing rice straw in flooded rice is considered a mitigation strategy that could theoretically reduce the GWP of emissions from rice by 45% (Wang et al. 2016). The benefits of complete straw removal on reducing emissions, however, are offset by reduced SOC sequestration, soil quality, and long-term yields. Maximum emission reductions and yield (and SOC deposition) may be best achieved by partial straw return/removal in most continuous rice systems (Romasanta et al. 2017). This balance can still increase SOC storage over time and provide adequate crop nutrients. Because straw decomposition rates, and thus emissions, depend on climate, cropping system, and soil type, these factors can help determine the appropriate percentage of straw to return. Generally, soils that are well-drained or have low SOC with aerobic periods benefit from increased straw return to maximize SOC sequestration and increase yields with minimal CH_4 emissions, i.e., the percentage of straw returned should be approximately proportional to the percentage of time under aerobic conditions (Monteleone et al. 2015).

Controlling the aerobic condition of paddy soil is primarily achieved by irrigation management. The use of non-flooded, aerobic periods to reduce CH_4 from organic matter decomposition in rice is a well-established mitigation strategy called alternate wetting and drying (AWD) that can reduce emissions in lowland irrigated rice by 48% on average (IRRI 2016). AWD will be an increasingly important strategy to mitigate future emissions of CH_4 as expanding combine harvester use promotes straw incorporation. Reduced flooding can also be achieved with the use of laser land-levelling, dry direct-seeded rice, and short-duration rice varieties. These methods are well established water-saving practices described in previous studies (Monteleone et al. 2015; Bouman et al. 2007). Reduced flooding affects emissions by shifting from anaerobic to aerobic microbial respiration to produce CO_2 in place of CH_4. Although CO_2 emissions increase under aerobic conditions, the effect on GWP is much lower than CH_4. Additionally, aerobic decomposition of residue improves SOM conversion to more stabilized forms of SOC that have a lower additive effect on CH_4 once flooded (Jiang et al. 2019).

Despite the benefits of aerobic regimes on emissions from rice straw, it comes with an increased risk of SOC loss compared to continuous flooding. Additionally, N_2O emissions may be significant during dry conditions— although N_20 emissions are largely an effect of fertilizer, as straw supplies only around 10% of N in intensive systems (Yadvinder-Singh et al. 2004; Eagle et al. 2001). In more aerobic rice systems, N_2O emissions can be mitigated by proper nutrient management, and SOC losses can be compensated for by increasing the rate of straw return.

9.2.2.2 Cropping System

As with irrigation management, the type of cropping system is an important factor in controlling soil conditions and emissions from rice straw. Because fallow conditions and upland crops mostly eliminate anaerobic conditions for CH_4 production, the emissions from aerobic decomposition (N_2O, CO_2) and loss in SOC can be significant. For example, SOC levels in a long-term rice–maize rotation at IRRI were 14% lower than that of continuous rice (Witt et al. 2000). For this reason, intensive rice–upland cropping systems may require complete straw return to the upland crop to prevent SOC depletion.

9.2.2.3 Tillage

Tillage type and timing can greatly affect emissions from straw returned to the field. When straw is chopped and incorporated into the soil at least 30 days before flooding, rice CH_4 emissions have been shown to be reduced by up to 80% (Launio et al. 2013; Kajiura et al. 2018). Reduction CH_4 emissions can be attributed to the increased aerobic decomposition of straw to stabilized SOM before flooding. Due to the additional benefits of early incorporation to planting and soil quality, it is considered a CSA priority for flooded rice. In fact, studies show early incorporation is one the most cost-effective, climate-smart rice straw management options (Launio et al. 2016).

When residue is removed, tillage has shown to increase emissions and reduce SOC in rice. A meta-analysis on 48 studies on continuous rice in China showed that no-till reduced the GWP from CO_2 and CH_4 by 20.4% when straw was removed, but had no significant effect when straw was returned (Feng et al. 2018; Huang et al. 2018).

In upland crops after rice, no-tillage with full straw returned is an established CSA strategy for many rice–upland environments (Grace et al. 2012). A study on marginal abatement costs suggest that no-till accounted for 70% of the cost-effective GHG mitigation potential in 2010 across non-rice crops (EPA 2013). The effects of no-till and straw mulching on yield, GHG emissions, and soil quality are most pronounced in rainfed, light textured soils. In fact, no-till for the rice–wheat rotation is credited as one of the greatest resource-saving technologies for the Indo-Gangetic Plains (Erenstein 2009; Zandstra 1982). Tillage is shown to stimulate mineralization and oxidation of SOM in aerobic soils, causing a reduction in SOC and increase in N_2O emissions. These effects have been established in many meta-analyses (Zhao et al. 2015; Feng et al. 2018; Lu 2015; Liu et al. 2014). Therefore, the optimal tillage management for CSA in rice–upland systems is often complete straw returned as mulch with no-till in the upland crop followed by early residue incorporation or removal before flooded rice.

9.2.2.4 Soil Type

Emissions and mitigation from rice straw management are highly dependent on soil type (Badagliacca et al. 2017). A meta-analysis of GHGE studies across Japan showed that CH_4 emissions significantly varied by soil type by as much as 200% (Kajiura et al. 2018). Still, the soil properties that stimulate CH_4 emissions from straw incorporation are not well understood. Conditions known to stimulate methanogenesis are a soil redox potential below −200 mV and neutral pH. It can be assumed that the variability in CH_4 production by soil type may be related to differences in soil nutrients. Some studies suggest that high levels of ammonia and sulfates are known to inhibit methanogenesis (Sánchez et al. 2015).

The ability of straw incorporation to improve SOC sequestration is also affected by soil type. Generally, soils which have been depleted of SOC and contain high clay or oxygen-reduced conditions can store more C. It is estimated that returning crop residues to these soils along with proper CSA management could help sequester enough SOC to offset the current increase in emissions from all anthropogenic sources (White 2017).

9.2.2.5 Fertilizer

Studies suggest there is a significant interaction effect of rice straw management and fertilizer on GHGEs. Yet, the degree of this effect is complex and thus difficult to form conclusions on management recommendations. N_2O emissions from the application of organic and inorganic fertilizers are, however, an important topic as they represent 5% of global anthropogenic emissions (IPCC 2007). Although N_2O is considered negligible during most rice production, which is flooded or kept saturated, trends towards more aerobic rice systems due to water limitations and increasing upland crop rotation make N_2O a concern. A meta-analysis on 112 assessments showed that straw incorporation can reduce N_2O emissions from fertilizer by 27% in rice, although straw incorporation alone generally increased N_2O due to the inherent N content of straw (Shan and Yan 2013). There is also evidence that CH_4 emissions from straw incorporation are affected by fertilizer. A meta-analysis of 155 data pairs showed that N fertilizer stimulated CH_4 emissions in 64% of cases and the stimulatory effect of N fertilizer on CH_4 was two to threefold greater with urea than with ammonium sulphate (Banger et al. 2012).

9.3 Off-Field Straw Management Effects on GHGEs

9.3.1 Composting

Straw composting with manure can be an effective option to reduce CH_4 emissions associated with in-field straw incorporation along with CH_4 and N_2O emissions from manure management. Manure management accounts for 11% of global agricultural emissions, thus is an equally important GHG source as lowland rice. Emissions from manure are mainly in the form of CH_4 from anaerobic settling ponds (23%) and N_2O from manure applied to soils and dry storage (77%) (FAO 2017).

Aerobic composting is an effective method to reduce methanogenesis of CH_4 from anaerobic manure storage in settling ponds. Studies suggest aerated manure with straw can reduce CH_4 emissions up to 90% compared to anaerobic storage (Petersen et al. 2013). The effects of composting on N_2O emissions from manure are, however, more complex than CH_4. N_2O is emitted indirectly from manure mainly by NH_3 volatilization, which converts to N_2O in the atmosphere. Smaller, but additional N losses can occur from NO_3 leaching/erosion, which also convert to N_2O. Improper field application of manure or composting can cause an almost 100% loss of manure N to the atmosphere affecting both GHGEs and N supply value if used for fertilizer. This often occurs when manure is applied to soils with high pH and low CEC, and without injection/incorporation. In this scenario, composting manure with rice straw could provide substantial emissions mitigation.

Rice straw is an ideal bulking agent for manure compost due to its high C:N ratio, which can help maintain the ideal 25:1 of the compost. This C:N ratio maximizes N immobilization and substrate adsorption, which minimizes losses by volatilization and leaching. N losses from proper composting may be as low as 13% of the original feedstock N (Chadwick et al. 2011). The opportunity to mitigate N_2O from composting, however, may be fairly small given many farms can avoid 100% N loss by injecting/incorporating manure or applying it directly to soils with high CEC, clay, or low pH. In this case, the mitigation opportunity of straw/manure compost may be primarily through avoiding CH_4 emissions from anaerobic manure storage and in-field rice straw incorporation, along with the potential indirect abatement of emissions from N fertilizer production (Chen et al. 2011). An additional, yet understudied, effect of rice straw composting vs. in-field incorporation may come from increased SOC sequestration. Although studies are limited, some suggest composting increases the stabilized fraction of SOC and sequesters more carbon compared to in-field aerobic decomposition of residue (Spaccini and Piccolo 2017).

The added step of producing mushrooms from straw compost could theoretically reduce N_2O emissions further by increasing N immobilization through mushroom nutrient uptake, although this has not been established. Studies do suggest that in-field emissions of CH_4 can be substantially mitigated by incorporating spent mushroom compost to the field in place of fresh rice straw. One study in the Philippines estimated CH_4 emissions from mushroom production at only 73 g CH_4 t^{-1} of straw (dry weight) compared to the IPCC default emission factor of 4 kg CH_4 t^{-1} for straw manure compost (Truc 2011). Arai et al. (2015) also found that the total GWP in straw-mushroom cultivation is 12.5% lower than straw burning.

9.3.2 Biochar

Like compost, biochar can mitigate the CH_4 emissions associated with fresh straw incorporation by providing an off-field use for straw. The total mitigation potential of biochar, however, extends beyond compost due to its ability to improve sequestration by converting straw to a more stabilized form of C (Yin et al. 2014). Studies on C cycling of crop residue suggest that incorporation and composting lose 80–90% of the initial carbon as CO_2 during decomposition in the first 5–10 years. In contrast, about 50% of the carbon can be captured as stable SOC when residue is converted to biochar (Lehmann et al. 2006)

Biochar blended with manure/straw compost has also been shown to substantially reduce N losses during the composting process due to its effect on nutrient sorption. Like straw, biochar can increase the adsorption of N and prevent NH_3 volatilization and this effect from biochar can be many times greater than that of straw due its high adsorption capacity or CEC. Studies on compost showed total N losses could be reduced by 52% with the addition of biochar (Steiner et al. 2010).

When biochar is returned to the field, its effects on total GHGEs; however, are mixed—possibly due to the variable quality of biochar products and dynamic conditions of soil. A meta-analysis of 61 studies on biochar of various feedstocks showed that GHGEs in paddy rice were: −5% for CO_2, −20% for N_2O, but +19% for CH_4 ($P < 0.05$) with the addition of biochar (Song et al. 2016). Conversely, another meta-analysis of 42 studies showed that biochar reduced CH_4 in acidic soils (Jeffrey et al. 2016). A CH_4 reduction along with a 50–70% reduction in the total C footprint for rice production was also reported in a life cycle assessment study comparing open-field straw burning to straw biochar (Mohammadi et al. 2016). A meta-analysis of 29 studies comparing biochar effects among cropping systems showed that biochar reduced GHGI (yield-scaled emissions) by 41% in upland soils and 17% in paddy soils (Liu et al. 2019).

In light of those studies with large emissions reductions, some authors suggest biochar could potentially mitigate emissions of CO_2, CH_4, and N_2O by a maximum of 1.8 Pg CO_2 eq year^{-1} (12% of current anthropogenic CO_2 eq emissions; 1 Pg = 1 Gt), and total net emissions over the course of a century by 130 Pg CO_2 eq (Das et al. 2014). Theoretically, this makes biochar one of the top mitigation options for rice straw management. Still, more evidence is needed on the feasibility of biochar in CSA, especially as many studies suggest it is cost-prohibitive due to the large volume (around 6 t ha^{-1}) of biochar needed in-field to achieve mitigation.

9.4 Other Off-Field Practices and Effects on GHGEs

9.4.1 Mechanized Straw Collection

The use of combine harvesters for rice has expanded rapidly worldwide, and major producers such as Vietnam and Cambodia almost exclusively rely on them (Gummert et al. 2018). This has large implications for rice straw management and its associated indirect and direct effects on GHG emissions. Contrary to traditional harvesting systems that use threshers and pile straw for easy collection, combine harvesters spread rice straw on the field. This hampers manual collection, thus promoting straw incorporation and increased CH_4 emissions. Additionally, the added emissions from fuel consumption and machine production range around 60–165 kg CO_2 eq t^{-1} of collected straw (Nguyen et al. 2016).

9.4.2 Fodder

Enteric fermentation as CH_4 from livestock is the leading source of agricultural emissions and accounts for about 5.8% of total anthropogenic emissions (Gerber et al. 2013). The quality of ruminant feed has a significant effect on this emission intensity. Rice straw fodder, although used widely across Asia, is particularly inefficient as a ruminant feed. Its low digestibility equates to high yield-scaled CH_4 emissions compared to more high-quality fodder, such as cowpea straw (Hristov et al. 2013). In fact, rice straw as fodder has been shown to increase GWP 13% compared to straw burning (Launio et al. 2016). Because of the widespread use of rice straw as fodder, it can be assumed that its contribution to emissions from enteric fermentation is significant. Improving the digestibility of poor-quality fodder, such as rice straw, may be one of the most effective emissions mitigation strategies for livestock according to Gerber et al. (2013). Research suggests that the digestibility of rice straw could be improved by up to 20% by pretreatment methods, such as nutrients and inoculants (Sarnklong et al. 2010). In cattle, a 1% increase in straw digestibility equates to a 4% increase in growth rate and proportional drop in yield-scaled emissions.

9.4.3 *Bioenergy*

9.4.3.1 Straw Combustion for Thermal Bioenergy

Rice straw can serve as a low-cost and renewable fuel source for combustion power plants. According to LCA on the use of rice straw as thermal bioenergy in Thailand, emissions can be reduced by 1.79 kg CO_2 eq kWh^{-1} compared to coal power and 1.05 kg CO_2 eq kWh^{-1} compared to natural gas-based power generation. Delivand et al. (2011) found that substituting natural gas or coal fuels with rice straw fuels for power generation would result in a considerable fossil fuel savings and lower GHGEs. It was estimated that 0.378 tCO_2 eq t^{-1} straw and 0.683 tCO_2 eq t^{-1} straw could be avoided if rice straw substitutes natural gas or coal in the power generation sector, respectively.

9.4.3.2 Straw Anaerobic Digestion for CH_4 Bioenergy

Agricultural residues, such as rice straw, offer a valuable alternative feedstock for biogas production since they contain a considerable amount of carbon that is beneficial for anaerobic codigestion with animal manure (Mussoline et al. 2012). Anaerobic digestion (see more details in Chap. 5) is a biological process that can degrade waste organic material by the concerted action of a wide range of microorganisms in the absence of oxygen. The process converts a large portion of rice straw into biogas, which is typically a mixture of methane (60%) and carbon dioxide (40%). If captured, biogas can be utilized as a clean fuel for heat and power generation. In principle, anaerobic digestion is an attractive option for mitigating the CH_4 associated with straw incorporation. However, in actual practice, particularly for small-scale anaerobic digestion, the technology has not proven efficient enough to be the most feasible mitigation strategy. Improving the technology to reduce leakage and match the digester capacity to biogas use in small-scale applications may be required to be a viable mitigation option.

Regarding the use of rice straw for bio-ethanol production, a review by Cheng and Timilsina (2011) reported that all advanced biofuel technologies have the advantage of producing fuels with almost zero or very little net emissions to the atmosphere.

9.5 Conclusions and Recommendations

Lowland rice contributes 10% of the global agricultural GHGEs due to CH_4 production from anaerobic decomposition of organic material. Straw management is therefore a key factor for controlling global agricultural emissions. Incorporating rice straw under flooded conditions leads to high CH_4 emissions. Burning, although a standard practice with lower GHGEs than incorporating, is not considered a CSA option due to its negative effect on soil nutrients, SOC, and air pollution. Water management through AWD is a major GHG mitigation strategy that can reduce 48% of the CH_4 and thus is an effective method to reduce emissions when straw is incorporated under flooded conditions. AWD in combination with early incorporation can further reduce CH_4 emissions by 80%. The rate of straw incorporation to achieve CSA, however, is highly dependent on environment. Rice–upland crop rotations or rice systems with prolonged fallow periods benefit from greater rates of straw incorporation due to losses in SOC. High rates of straw incorporation under aerobic conditions can sequester SOC with a minimal increase in emissions compared to incorporation under flooded conditions. Practices that optimize SOC sequestration while minimizing emissions, such as early straw incorporation with AWD water management could be an important step towards carbon neutral rice systems.

Off-field practices such as composting, biochar, and bioenergy offer potentially larger mitigation opportunities than in-field practices. Composting, for example, can mitigate both emissions associated with fresh straw incorporation and those associated with livestock manure and fertilizer use. The combination of biochar and compost can further enhance mitigation. Although effective, off-field technologies may be limited due to the added costs of straw transport, capital equipment and labor.

Depending on site-specific conditions related to economics, climate, soil type, and infrastructure, a combination of off-field and in-field straw management practices is needed to reduce emissions from rice production. More holistic and cross-sectoral studies, e.g., through life-cycle assessment, are needed to determine the full GHG budget of certain site-specific straw management options. Additionally, MACC and CBA studies would be important to develop clear technical and policy recommendations that also consider the economics of CSA and straw management.

References

Arai T, Takaya T, Ito Y, Hayakawa K, Tshima S, Shibuya C, Nomura M, Yoshimi N, Hibayama M, Yasuda Y (1998) Bronchial asthma induced by rice. Intern Med 37:98–101

Arai H, Hosen Y, Pham Hong VN, Thi NT, Huu CN, Inubushi K (2015) Greenhouse gas emissions from rice straw burning and straw-mushroom cultivation in a triple rice cropping system in the Mekong Delta. Soil Sci Plant Nutr 61:719–735

Badagliacca G, Ruisi P, Rees RM, Sergio S (2017) An assessment of factors controlling N_2O and CO_2 emissions from crop residues using different measurement approaches. Biol Fertil Soils 53:547

Banger K, Tian H, Lu C (2012) Do nitrogen fertilizers stimulate or inhibit methane emissions from rice fields? Glob Chang Biol 18:3259–3267

Bouman BAM, Lampayan RM, Tuong TP (2007) Water Management in Irrigated Rice: coping with water scarcity. Intl Rice Res Inst, Los Baños, p 54

Chadwick D, Sommer SG, Thorman R, Fangueiro D, Cardenas L, Amon B, Misselbrook T (2011) Manure management: implications for greenhouse gas emissions. Anim Feed Sci Technol 166–167:514–531

Chen J, Gong Y, Wang S, Guan B, Balkovic J, Kraxner F (2019) To burn or retain crop residues on croplands? An integrated analysis of crop residue management in China. Sci Total Environ 662:141–150

Chen R, Lin x WY, Hu J (2011) Mitigating methane emissions from irrigated paddy fields by application of aerobically composted livestock manures in eastern China. Soil Use Manag 27:103–109

Cheng JJ, Timilsina GR (2011) Status and barriers of advanced biofuel technologies: a review. Renew Energy 36:3541–3549

Das S, Avasthe R, Singh R, Babu S (2014) Biochar as carbon negative in carbon credit under changing climate (vol 107)

Delivand MK, Barz M, Gheewala SH (2011) Logistics cost analysis of rice straw for biomass power generation in Thailand. Energy 36:1435–1441

Eagle AJ, Bird BA, Horwath WR, Linquist BA, Brouder SM, Hill JE, Van Kessel C (2001) Rice yield and nitrogen utilization efficiency under alternative straw management practices. Agron J 92:1096–1103

Erenstein O (2009) Zero tillage in the rice-wheat systems of the Indo-Gangetic plains: A review of impacts and sustainability implications. IFPRI Discussion Paper 00916. International Food Policy Research Institute, Washington

EPA (US Environmental Protection Agency) (2013) Global mitigation of Non-CO_2 greenhouse gases, 2010–2030.Washington, DC. https://www.epa.gov/sites/production/files/2016-06/documents/mac_report_2013.pdf

FAO (Food and Agriculture Organization) (2015) Estimating greenhouse gas emissions in agriculture: a manual to address data requirements for developing countries. Food and Agriculture Organization of the United Nations, Rome

FAO (Food and Agriculture Organization) (2017) Global Livestock Environmental Assessment Model (GLEAM) http://www.fao.org/gleam/results/en/

Feng J, Li F, Zhou X, Xu C, Ji L, Chen Z, Fuping F (2018) Impact of agronomy practices on the effects of reduced tillage systems on CH_4 and N_2O emissions from agricultural fields: a global meta-analysis. PLoS One 13(5):e0196703

Fileccia T, Guadagni M, Hovhera V, Bernoux M (2014) Ukraine: soil fertility to strengthen climate resilience preliminary assessment of the potential benefits of conservation agriculture. Food and Agriculture Organization of the United Nations, Rome

Gerber PJ, Steinfeld H, Henderson B, Mottet A, Opio C, Dijkman J, Falcucci A, Tempio G (2013) Tackling climate change through livestock: a global assessment of emissions and mitigation opportunities. Food and Agriculture Organization of the United Nations, Rome

Grace PR, Antle J, Aggarwal PK, Ogle S, Paustian K, Basso B (2012) Soil carbon sequestration and associated economic costs for farming systems of the indo-Gangetic plain: a meta-analysis. Agric Ecosyst Environ 146:137–146

Gullett B, Touati A (2003) PCDD/F emissions from burning wheat and Rice field residue. Atmos Environ 37:4893–4899

Gummert M, Quilty J, Hung NV, Vial L (2018) Engineering and management of rice harvesting. In: Pan Zhongli Khir R Advances in science & engineering of rice Destech Publication Pennsylvania

Hristov AN, Oh J, Lee C, Meinen R, Montes F, Ott T, Firkins J, Rotz A, Dell C, Adesogan A, Yan W, Tricarico J, Kebreab E, Waghorn G, Dijkstra J, Oosting S (2013) Mitigation of greenhouse gas emissions in livestock

Huang Y, Ren W, Wang L, Hui D, Grove JH, Yang X, Tao B, Goff B (2018) Greenhouse gas emissions and crop yield in no-tillage systems: a meta-analysis. Agric Ecosyst Environ Appl Soil Ecol 268:144–153

IRRI (International Rice Research Instititute) (2016) Overview of AWD. http://books.irri.org/AWD_brochure.pdf

IPCC (Intergovernmental Panel on Climate Change) (2006) IPCC guidelines for national greenhouse gas inventories. Prepared by the National Greenhouse Gas Inventories Programme, Eggleston HS, Buendia L, Miwa K, Ngara T, Tanabe K (eds) IGES: Japan

IPCC (Intergovernmental Panel on Climate Change) (2007) Climate change 2007: the physical science basis summary for policymakers. Contribution of working group I to the fourth assessment report of the intergovernmental panel on climate change. In: Solomon S, Qin D, Manning M, Chen Z, Marquis M, Averyt KB, Tignor M, Miller HL, eds. Cambridge University Press, New York

Jeffrey S, Verheijen FGA, Kammann C, Abalos D (2016) Biochar effects on methane emissions from soils: a meta-analysis. Soil Biol Biogeochem 101:251–258

Jiang Y, Qian H, Huang S, Zhang X, Wang L, Zhang L, Shen M, Xiao X, Chen F, Zhang H, Lu C, Li C, Zhang J, Deng A, van Groenigen KJ, Zhang W (2019) Acclimation of methane emissions from rice paddy fields to straw addition. Sci Adv 5:eaau9038

Kajiura M, Minamikawa K, Tokida T, Shirato Y, Wagai R (2018) Methane and nitrous oxide emissions from paddy fields in Japan: an assessment of controlling factors using an intensive regional data set. Agric Ecosyst Environ 252:51–60

Kanokkanjana K, Cheewaphongphan P, Garivait S (2011) Black carbon emission from paddy field open burning in Thailand. IPCBEE Proc 6:88–92

Lal R (2004) Soil carbon sequestration impacts on global climate change and food security. Science 304:1623–1627

Launio CC, Asis CA, Manalili RG, Javier EF (2013) Economic analysis of rice straw management alternatives and understanding farmers' choices. EEPSEA Research Report rr2013031, Economy and Environment Program for Southeast Asia. https://ideas.repec.org/p/eep/report/rr2013031.html

Launio CC, Asis CA, Manalili RG, Javier EF (2016) Cost-effectiveness analysis of farmers' rice straw management practices considering CH_4 and N_2O emissions. J Environ Manag 183:245–252

Lehmann J, Gaunt J, Rondon M (2006) Bio-char sequestration in terrestrial ecosystems: a review. Mitig Adapt Strat Glob Change 11:403–427

Lin LF, Lee WJ, Li HW, Wang MS, Chang-Chien GP (2007) Characterization and inventory of PCDD/F emissions from coal-fired power plants and other sources in Taiwan. Chemosphere 68:1642–1649

Liu C, Lu M, Cui J, Li B, Fang C (2014) Effects of straw carbon input on carbon dynamics in agricultural soils: a meta-analysis. Glob Chang Biol 20:1366–1381

Liu X, Mao P, Li L, Ma J (2019) Impact of biochar application on yield-scaled greenhouse gas intensity: a meta-analysis. Sci Total Environ 656:969–976

Lu F (2015) How can straw incorporation management impact on soil carbon storage? A meta-analysis. Mitig Adapt Strateg Glob Change 20:1545

Lu F, Wang X, Han B, Ouyang Z, Duan X, Zheng H (2010) Net mitigation potential of straw return to Chinese cropland: estimation with a full greenhouse gas budget model. Ecol Appl 20(3):634–647

Mohammadi A, Cowie A, Anh Mai TL, de la Rosa RA, Kristiansen P, Brandão M, Joseph S (2016) Biochar use for climate-change mitigation in rice cropping systems. J Clean Prod 116:61–70

Monteleone M, Garofalo P, Cammerino ARB, Libutti A (2015) Cereal straw management: a trade-off between energy and agronomic fate. Ital J Agron 10(2):59–66

Mosier AR, Halvorson AD, Reule CA, Liu XJ (2006) Net global warming potential and greenhouse gas intensity in irrigated cropping systems in northeastern Colorado. J Environ Qual 35:1584–1598

Mussoline W, Esposito G, Giordano A, Lens P (2012) Anaerobic digestion of rice straw: a review. Crit Rev Environ Sci Technol 43(9):895–915

Nguyen H, Nguyen C, Tran T, Hau H, Nguyen N, Gummert M (2016) Energy efficiency, greenhouse gas emissions, and cost of rice straw collection in the mekong river delta of Vietnam. Field Crop Res 198:16–22

Petersen SO, Blamchard M, Chadwick D, Del Prado A, Edourd N, Mosquera J, Sommer SG (2013) Manure management for greenhouse gas mitigation. Animal 7 (s2):266–282

Robertson GP, Grace PR (2004) Greenhouse gas fluxes in tropical and temperate agriculture: the need for a full-cost accounting of global warming potentials. Environ Dev Sustain 6:51–63

Romasanta RR, Sander OB, Gaihre YK, Alberto MC, Gummert M, Quilty J, Nguyen VH, Castalone AG, Balingbing C, Sandro K, Correa T, Wassmann R (2017) How does burning of rice straw affect CH_4 and N_2O emissions? A comparative experiment of different on-field-straw management practices. Agric Ecosyst Environ 239:143–154

Sánchez A, Artola A, Font X, Gea T, Barrena R, Gabriel D, Monedero MAS, Roig A, Cayuela ML, Mondin C (2015) Greenhouse Gas from Organic Waste Composting: Emissions and Measurement. In: Lichtfouse E, Schwarzbauer J, Robert D (eds) CO2 sequestration, biofuels, and depollution. Environmental chemistry for a sustainable world. vol 5. Springer, Berlin

Sarnklong C, Cone JW, Pellikaan W, Hendriks WH (2010) Utilization of rice straw and different treatments to improve its feed value for ruminants: a review. Asian-Aust J Anim Sci 23(5):680–692

Shan J, Yan XY (2013) Effects of crop residue returning on nitrous oxide emissions in agricultural soils. Atmos Environ 71:170–175

Song X, Pan G, Zhang C, Zhang L, Wang H (2016) Effects of biochar application on fluxes of three biogenic greenhouse gases: a meta-analysis. Ecosyst Health Sustain 2:2. https://doi.org/10.1002/ehs2.1202

Spaccini R, Piccolo A (2017) Soil organic carbon stabilization in compost amended soils. Global Symposium on Soil Organic Carbon. 21–23 March 2017. Rome, Italy

Steiner C, Das KC, Melear N, Lakly D (2010) Reducing nitrogen loss during poultry litter composting using biochar. J Environ Qual 39:1236–1242. https://doi.org/10.2134/jeq2009.0337

Tipayarom D, Kim Oanh NT (2007) Effects from open rice straw burning emission on air quality in the Bangkok metropolitan region. Sci Asia 33:339–345

Torigoe K, Hasegawa S, Numata O, Yazaki S, Matsumaga M, Boku N, Hiura M, Ino H (2000) Influence of emission from rice straw burning on bronchial asthma in children. Pediatr Int 42:143–150

Truc N (2011) Comparative assessment of using rice straw for rapid composting and straw mushroom production in mitigation greenhouse gas emissions in Mekong Delta, Vietnam and Central Luzon, Philippines. Unpublished PhD thesis. University of the Philippines Los Baños

UNFCC (United Nations Framework Convention on Climate Change) (2019). http://di.unfccc.int/ghg_profile_non_annex1

Wang W, Wu X, Chen A, Xie X, Wang Y, Yin C (2016) Mitigating effects of ex situ application of rice straw on CH_4 and N_2O emissions from paddy-upland coexisting system. Sci Rep 6:37402

White J (2017).Gathering momentum around soil carbon sequestration. https://ccafs.cgiar.org/blog/gathering-momentum-around-soil-carbon-sequestration#.XOSZS1IzbIU%20IPCC,%202007

Witt C, Cassman KG, Olk DC, Biker U, Liboon SP, Samson MI, Ottow JCG (2000) Crop rotation and residue management effects on carbon sequestration, nitrogen cycling and productivity of irrigated rice systems. Plant Soil 225:263

Yadvinder-Singh B-S, Ladha JK, Khind CS, Khera TS, Bueno CS (2004) Effects of residue decomposition on productivity and soil fertility in fice–wheat rotation. Soil Sci Soc Am J 68:854–864

Yin YF, He XH, Gao R, Ma HL, Yang YS (2014) Effects of rice straw and its biochar addition on soil labile carbon and soil organic carbon. J Integr Agric 13:491–498

Zandstra HG (1982) Effect of soil moisture and texture on the growth of upland crops after wetland rice. In: Report of a workshop on cropping systems research in Asia. IRRI, Los Baños, Philippines

Zhao X, Zhang R, Xue JF, Pu C, Zhang XQ, Liu SL, Chen F, Lal R, Zhang HL (2015) Management-induced changes to soil organic carbon in China: a meta-analysis. Adv Agron 134:1–50

Chapter 10
Life Cycle Assessment Applied in Rice Production and Residue Management

Nguyen Van Hung, Maria Victoria Migo, Reianne Quilloy, Pauline Chivenge, and Martin Gummert

Abstract Rice production can be carried out using a wide set of cultivation techniques. Different land preparation, crop establishment, crop care, harvesting, and straw management techniques lead to different environmental impacts. Life-cycle assessment (LCA) is a reliable tool for assessing the environmental load of agricultural processes and can be used to compute or simulate energy balance and environmental impact categories such as climate change, ozone depletion, terrestrial acidification, freshwater eutrophication, and marine eutrophication. This chapter comprises the following sections: (1) LCA overview and application in agriculture, (2) case studies of LCA to identify the best rice straw management practices, and (3) summary and suggestions for further applications.

Keywords Life cycle assessment · LCA · Impact assessment · Energy balance · GHG emissions balance

10.1 Introduction

Crop production (during the crop's life cycle) is done using a wide set of cultivation techniques and crop management, harvest, and postharvest procedures. The processes and relevant inputs and outputs, which are part of the life-cycle process, are shown in Fig. 10.1. Rice production is typically characterized by three major phases: preplanting, plant growth, and postproduction.

N. V. Hung (✉) · R. Quilloy · P. Chivenge · M. Gummert
International Rice Research Institute (IRRI), Los Baños, Laguna, Philippines
e-mail: hung.nguyen@irri.org; r.quilloy@irri.org; p.chivenge@irri.org; m.gummert@irri.org

M. V. Migo
Department of Chemical Engineering, College of Engineering and Agro-Industrial Technology, University of the Philippines Los Baños (UPLB), Los Baños, Laguna, Philippines
e-mail: mpmigo@up.edu.ph

M. Gummert et al. (eds.), *Sustainable Rice Straw Management*,
https://doi.org/10.1007/978-3-030-32373-8_10

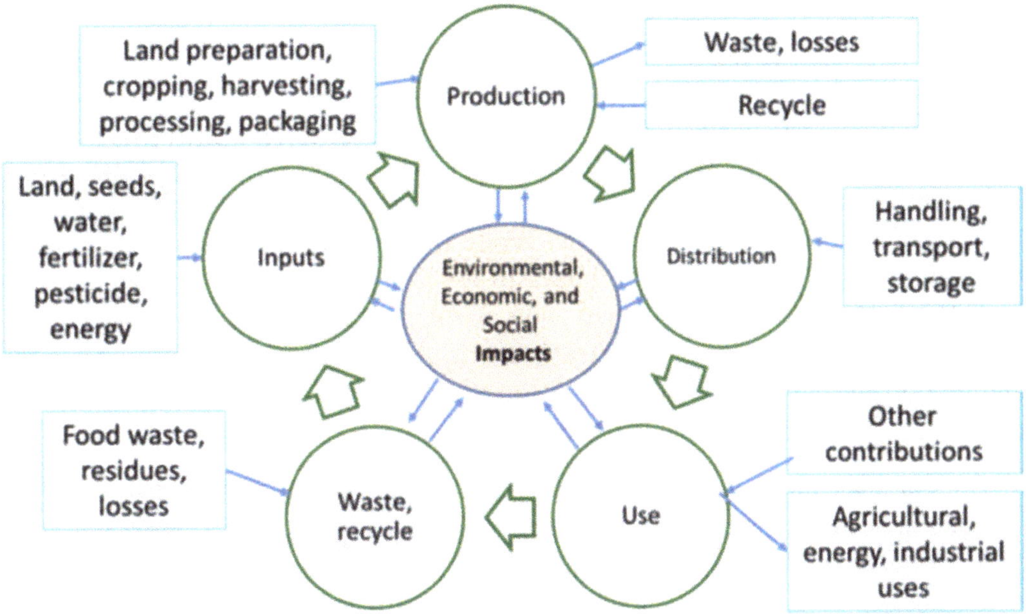

Fig. 10.1 Rice production life cycle

In preplanting, the variety is chosen and the field is prepared for planting. Land preparation consists of plowing or overturning the soil, harrowing or breaking the soil into smaller masses, and leveling the field. Equipment, which can be used in mechanized land preparation, include the power tiller, moldboard plow, hydrotiller, and rotovator. If herbicide is applied to kill weeds, the land is irrigated from 2 to 3 days after its application.

In the plant growth phase, rice is either directly seeded or transplanted. Fertilizers, pesticides, and herbicides are applied and, traditionally, the land is continuously flooded since rice is very sensitive to water shortages. Improved water management options, such as alternate wetting and drying (AWD), have been developed recently to conserve water and reduce greenhouse gas emissions (Carrijo et al. 2017; Linquist et al. 2015). Snails, which are persistent pests during this period, can be managed by manual removal or chemical control. During the growth phase, it is also important to apply N-P-K fertilizers and pesticides and herbicides (when necessary) to enhance yield. In addition, it is imperative to manage the water level to sustain the crop and control weeds.

Postproduction activities include harvesting, drying, storage, milling, and processing. Harvest operations include reaping, threshing, cleaning, hauling, field drying, piling, and bagging. Harvesting may involve traditional manual labor for all steps, semi-mechanical using a machine thresher, or full mechanization using a combine harvester. After combine harvesting, rice straw is left on the field and can either be collected, burned, or left to decompose (on the surface or after soil incorporation). When manually harvested, the rice stalks are cut and moved to a central threshing area, leaving behind uncut rice straw stubble, which can be burned or incorporated into the soil. Loose rice straw and stubble are considered byproducts. The paddy is then dried (traditional sun drying, solar drying, or mechanical drying) to reduce the moisture content to about 14%, which helps to prevent grain discoloration, mold formation, and insect attack. The paddy can be stored or milled to remove the husks and bran layers revealing the edible white kernels. The rice husks and bran are considered to be byproduct as well.

Life cycle assessment (LCA) is a tool for the analysis of the energy balance and environmental impacts of a process from cradle to grave, beginning with the gathering of raw materials from the earth to create the product and ending at the point when all materials are returned to the earth (US EPA 2006). LCA can be applied to compute or simulate energy balance and environmental impact categories, such as climate change, ozone depletion, terrestrial acidification, freshwater eutrophication, and marine eutrophication. LCA can also be applied for crop production and agricultural systems for the following purposes:

- Comparative analyses and identification of the best options among different production systems, practices, technologies based on some specific economic and environmental factors;
- Production process improvement, product development, and promotion; and
- Strategic planning and decision support.

10.2 LCA Framework

The LCA framework (Fig. 10.2) based on the ISO 14040 (Guinée et al. 2001) includes the following main components: (1) definition of goal, scope, and function unit; (2) inventory analysis; (3) impact assessment; and (4) interpretation. According to ISO 14040, the goal should contain an unambiguous description of the LCA's application and intended audiences as well as the reasons for conducting the study.

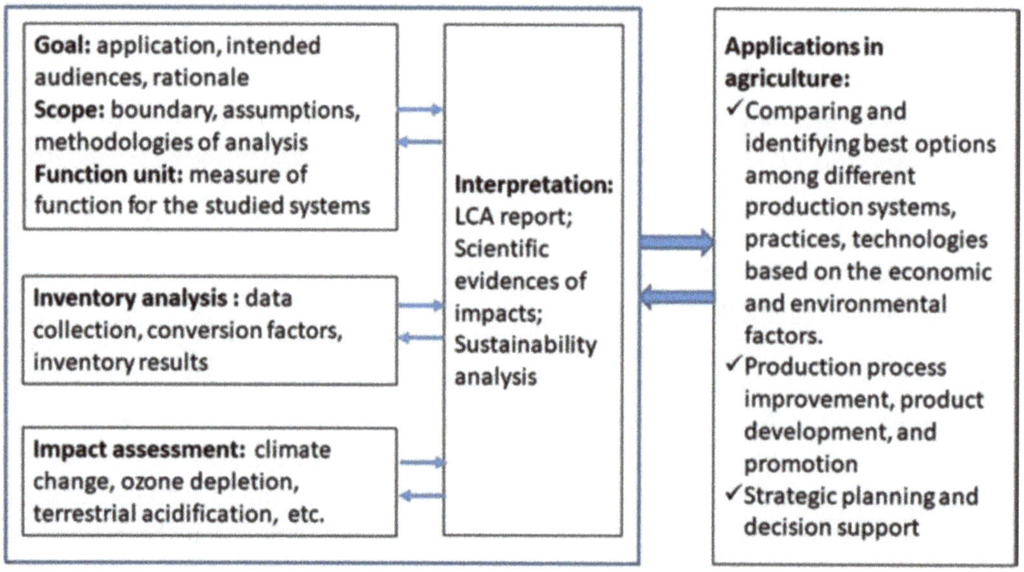

Fig. 10.2 LCA framework and applications

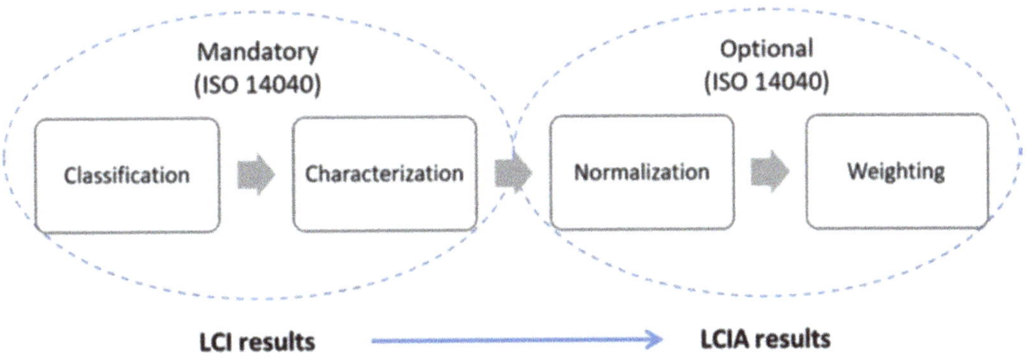

Fig. 10.3 LCIA steps (ISO 14040)

On the other hand, the scope should describe the most important methodological assumptions and limitations (PRé Product Ecology Consultants 2013). At this stage of the LCA, the functional unit or comparison basis is also defined. An example in rice science is the most common functional units of per kg or per ton of grain harvested. After defining the goal, scope, and functional unit, it is suggested to specify a system boundary that will determine which processes will be included in the LCA.

The life cycle inventory (LCI) involves listing all inputs and outputs and collecting data related to unit processes within the system boundary (Guinée 2004). In rice production, one input is the seed and one output is the grain. Thus, data collection on grain yield is necessary. The product of this step is called the LCI result. In some cases, systems have multi-functionality, as with rice production where there are two products, paddy and straw. Such cases require the LCA practitioner to allocate the inputs and outputs between the two products. However, according to the ISO 14040 standard, such allocation should be avoided as much as possible (Agri-footprint 2015). If allocation is unavoidable, the system inputs and outputs can be divided across mass allocation, gross energy allocation, and/or economic allocation (Agri-footprint 2015). Mass allocation is based on the mass or dry matter of the products. Gross energy allocation is based on the nutritional feed material list. Economic allocation is based on the prices of the products.

The next step, life cycle impact assessment (LCIA), is based on the LCI results as shown in Fig. 10.3. This analysis involves classification, characterization, normalization, and weighting, which are calculated based on ISO 14040. According to ISO 14040, classification and characterization are mandatory while normalization and weighting are optional.

10.2.1 Classification

Classification is a step in the identification of impact categories from LCI results (Fig. 10.4). LCI results include data on all inputs, such as land use, water, fuel consumption, labor, fertilizer, pesticides, and insecticides, and other direct emissions, such as CH_4, N_2O, CO_2, particulate matter ($PM_{2.5}$ and PM_{10}), etc. The calculated

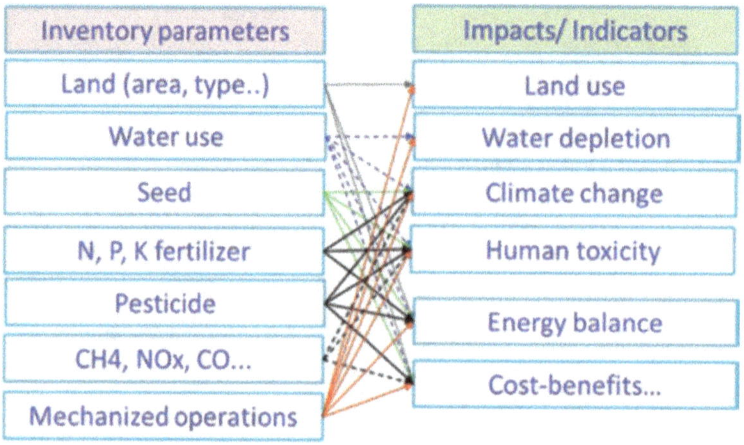

Fig. 10.4 Example of LCIA classification based on LCI results

results are classified into the impact categories, such as land use, water depletion, climate change, etc.

The results of different impacts are then translated into mass of CO_2 equivalent (for GWP or global warming potential), 1,4 dichlorobenzene-equivalent (for human toxicity), etc. This step is called characterization.

LCA indicators are measurable representations of an impact category. Description of the LCA indicators could be found in ISO 14040 and CML-1992 and -2001 (Jeroen et al. 2001; Budavari et al. 2011; De-Schryver et al. 2009; European Commission Joint Research Centre 2009; Guinée et al. 2001). Global warming indicators are presented in IPCC (2006). SIMAPRO is just one of many available LCA softwares or tools that incorporate indicators (SIMAPRO 2017). An overview of common indicator categories follows, particularly for rice production.

(a) ***Resources***

- Depletion of abiotic resources: the depletion of nonliving natural resources such as soil nutrients, etc. It can be described as the fraction of resource extracted over the recoverable reserves of that source. It is expressed in kg, m^3, or MJ $year^{-1}$.
- Cumulative energy demand: the total energy demand for production, including direct and indirect energy inputs. Direct energy inputs for crop production usually include agronomic inputs, fuel consumption, labor, etc.; while indirect energy inputs include the energy for production used by machines and other related infrastructures. The value is expressed in MJ.
- Water consumption: the impacts of water shortages due to groundwater extraction, expressed in m^3.
- Land use: highly relevant for crop production and alternative options for using land, expressed in area units (e.g., ha).

(b) ***Air Pollution***

- Global warming potential: the greenhouse effect instigated by the emissions of crop production from human activities. Greenhouse gas emissions

(GHGEs), such as N_2O and CH_4, intensifies the heat radiation absorption of the earth's atmosphere resulting to increasing surface temperatures. It is expressed in terms of mass (e.g., kgCO_2 equivalents).

- Ozone depletion potential: the thinning of the ozone layer in the stratosphere due to emissions from human activities which causes a potential damage to human health, ecosystems, biochemical cycles, and materials. It is described as the ratio between the amount of ozone destroyed by a unit of a substance and a reference substance, which is usually *Trichlorofluoromethane* (CFC-11). It is expressed in kg-CFC-11 equivalents.
- Acidification potential: the acidity of water and soil systems can be increased due to acid deposition from the atmosphere, mainly in the form of rain. Sulfur dioxide (SO_2), ammonia (NH_3) released through volatilization, and nitrogen oxides (NOx) emitted by combustion processes (such as burning rice straw) causes "acid rain." It is expressed in kg-SO_2 equivalents.

(c) ***Water Pollution***

- Eutrophication potential: the increase of the concentration of nutrients, chiefly nitrogen (N) and phosphorus (P), in a body of water caused by the runoff of synthetic fertilizers from agricultural land or by the input of sewage or animal waste. It causes the reduction in species diversity and the over-population of a dominant species, which is usually algae—a phenomenon called "algal bloom". In turn, the increased production of dead biomass from algae consumes oxygen thru a degradation process, and depletes the oxygen in the water. It is expressed in phosphate (PO_4^{3-}) equivalents.
- Aquatic ecotoxicity: the impact on fresh water ecosystems as a result of emissions of toxic substances into air, water, and soil. It is expressed as 1,4-dichlorobenzene equivalents (1,4 DB-eq) per kg of emission.

(d) ***Soil Pollution***

- Terrestrial ecotoxicity: the impact of toxic substances released into terrestrial ecosystems. It is defined as the potential of terrestrial toxicity of each substance emitted into the air, water, and/or soil and expressed as 1,4 DB-eq per kg of emission.

(e) ***Damage, Health, and Biodiversity***

- Human toxicity potential: the impact on human health of toxic substances present in the environment. Human toxicity is identified as the overall impact of toxic substances into air, water, and soil, which are most vulnerable to pollution and contamination, such as carbon monoxide (CO), black carbon from straw burning, heavy metal loads in water and soil, etc. These toxic substances accumulate in the vegetables, fruits, meat, milk and other animal products which in turn are ingested by humans. It is expressed in kg of 1,4 DB-eq.
- Disability-adjusted life years (DALY): It is the total years of life lost by premature mortality and the lost of productive life due to incapacity (Goedkoop and Spriensma 2001). This indicator, expressed in DALY kg^{-1} of emission, determines amounts of heavy metals and carcinogenic substances.

10.2.2 Normalization and Weighting

Normalization is a process to calculate the magnitude of the results of the impact indicators, relative to some reference information. Normalized results of the characterized factors, such as for each person per year, are calculated by dividing the characterized results by the normalization factors, which are standardized in ISO 14044 (Guinée et al. 2001). Normalization factors are different for midpoint and endpoint impact categories. These are described in Budavari et al. (2011).

Weighting is aimed at expressing the impact results for each category in numerical factors. Weighting is based on value choices or votes. For instance, there are different votes for the importance of different categories, e.g., climate change and human toxicity, in different regions or countries. Weighting factors of some common impact categories are presented in Budavari et al. (2011) and Guinée et al. (2001).

Table 10.1 provides an example of the characterization, normalization, and weighting steps.

As shown in Table 10.1, normalized results indicate that burning 1 ton of rice straw contributes to 2.3% (0.0229) of climate change impact and 0.15% (0.0015) of human toxicology impact of an average person in a year. The total weighting result accounted for these two impact factors is 0.51, which is referred as an impact score to compare with other scenarios.

Of the different steps, a bulk of the work is done in the LCI and LCIA. An LCA tool, such as SIMAPRO software, aids in the calculations during the LCIA. One advantage of using an LCA tool is the availability of global and regional databases, which contain data that are impossible to measure during the scope of the study. An example in rice science is the amount of energy expended to make 1 kg of fertilizer. These data are required to calculate the amount of energy in producing 1 kg of paddy or straw and the LCA practitioner using the software usually refers to data in

Table 10.1 Characterization, normalization, and weighting, for example, of rice-straw burning

LCI results			
Factors	Emissions (kg t^{-1} of straw)	Climate change: GWP-100a (kg CO_2-eq)	Human toxicity potential (kg 1,4 DB-eq)
CH_4	4.5	X 30.5 = 137.3	–
N_2O	0.07	X 265 = 18.6	–
$PM_{2.5}$	10	–	X 0.82 = 8.2
PM_{10}	6	–	X 0.82 = 4.9
Characterized results		155.8 kg CO_2-eq	13.1 kg 1,4 DB-eq
Normalization factor (Budavari et al. 2011)		6803 kg CO_2-eq/person/year	8800 kg 1.4 DB-eq/person/year
Normalized results		0.0229 person-year	0.0015 person-year
Weight factor (Budavari et al. 2011)		21.6	8
Weighting results for the two factors		0.51	

the literature via the built-in libraries. Therefore, knowledge on the actual composition of the fertilizer is necessary to correctly select the appropriate data from the libraries. Another responsibility of the practitioner is to review the documentation that comes with the databases to check compatibility of the data.

Lastly, the interpretation step is done by making well-balanced conclusions and recommendations based on the LCIA (Guinée 2004). This is also where sensitivity and uncertainty analyses can be made.

10.3 Some Typical and Advanced Analyses in LCA

10.3.1 Analyzing Energy and GHGE Balances

Energy and GHGE balances can be analyzed in LCA and calculated based on the following steps and equations:

- Net = $\sum$ output − $\sum$ input
- Net balance factor = Net/ $\sum$ input
- Net energy: $NE = \sum_{i=1}^{n} \text{output energy}_i - \sum_{j=1}^{m} \text{input energy}_j$
- $Net\,\text{energy balance} = NE / \sum_{j=1}^{m} \text{input\_energy}_j$
- GHGE balance: Net GHGE = $\sum$ GHGE of products and avoided products − $\sum$ GHGE of productions (inputs and emissions)

In rice science, the input energy for the supply chain accounts for rice cultivation, harvesting, collection and transportation of products (e.g., paddy and rice straw), storage of products, and processing. For rice cultivation, the input should cover energy from rice seeds, fertilizers, pesticides, fuel consumption, machine production, and labor.

10.3.2 Sensitivity and Uncertainty Analyses

An analysis model with the correlation of inputs, simulation process, outputs, and feedbacks is shown in Fig. 10.5. Data distribution, error, and affecting scenarios of the input models will affect the output models. Sensitivity analysis measures the change in the outputs affected by the scenarios of actual events or assumed in the inputs. On the other hand, uncertainty analysis is used to describe the entire set of possible outcomes, together with their associated probabilities of occurrence. The Monte Carlo computational algorithm is a common method used in uncertainty analysis. These methods are incorporated in some LCA software, such as SIMAPRO (2017).

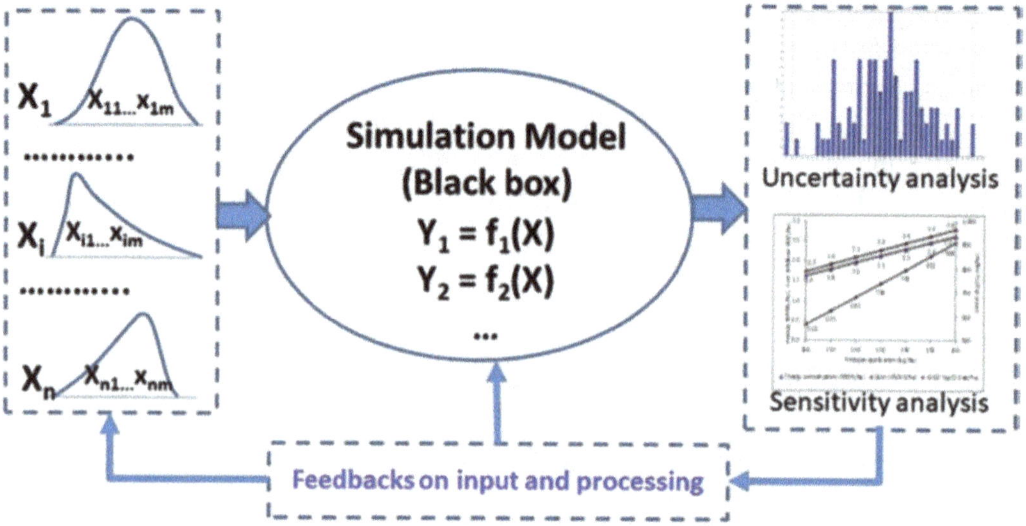

Fig. 10.5 Analysis model with the correlation of inputs, simulation process, outputs, and feedbacks

10.4 Case Study Using LCA and SIMAPRO for Rice Production

LCA research relevant to agricultural products or processes can be found in recent publications on rice straw bioenergy (Shie et al. 2011; Kami et al. 2012; Singh et al. 2013), rice straw biofuel and fertilizer (Silalertruksa and Gheewala 2013), power generation (Suramaythangkoor and Gheewalal 2011; Shafie et al. 2014), rice production (Brodt et al. 2014), rice straw management (Fusi et al. 2014), rice straw anaerobic digestion (Nguyen et al. 2016), and rice straw collection (Nguyen et al. 2017).

Here we present a case study on rice production for different rice straw management practices based on research conducted at IRRI from 2015 to 2016 (Nguyen et al. 2019). The goal and scope of the study were to compare environmental profiles, grain yield and quality, energy efficiency, and GHGEs of rice production at the IRRI farm during the 2015 wet season and 2016 dry season, followed by four different rice-straw management options: retaining the straw and incorporation, straw burning, partial straw removal, and complete straw removal through LCA (Fig. 10.6). The functional unit of the system is 1 ha of rice production and the impact results were translated to 1 t of rice based on the yield data.

For life-cycle inventory analysis, we measured operations and agricultural inputs of rice production on the IRRI farm (including grain yield and total biomass) during the 2015 wet and 2016 dry seasons. Soil sampling and analysis were also conducted. Inputs and outputs for each process in the system boundary (Fig. 10.6) were itemized. For LCIA, single scores on net energy and net GHGEs were evaluated. This step was done with the aid of SIMAPRO software. Data on net energy values

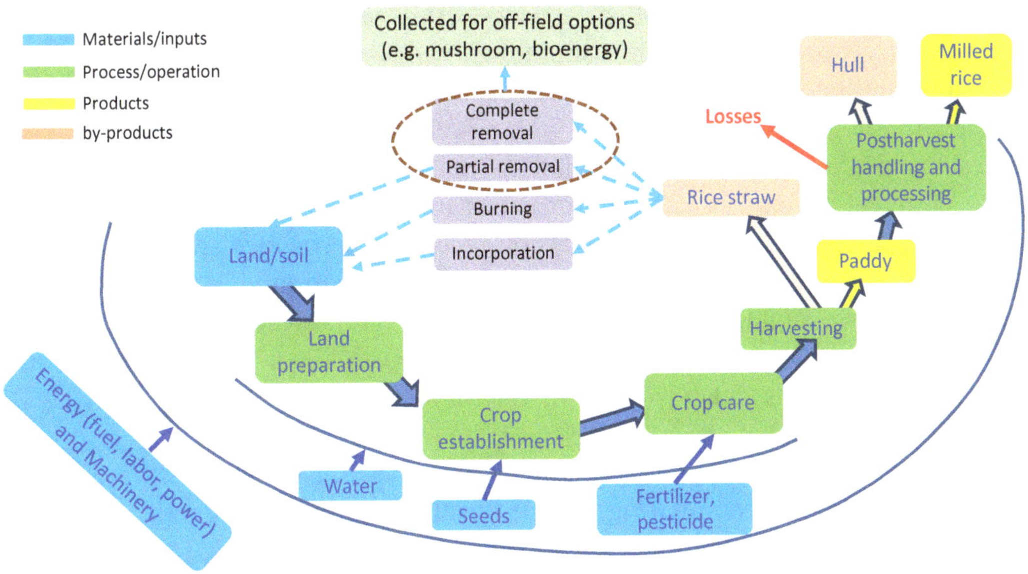

Fig. 10.6 Research boundary of LCA in rice production with different rice-straw management options. (Adapted from Nguyen et al. 2019)

Table 10.2 Energy and GHGE conversion factors of fuel, agronomic inputs, and products

	Energy			GHGE		
Parameters	Unit	Value	Sources	Unit	Value	Sources
Seeds	MJ kg^{-1}	30.1	a, b	$kgCO_2$-eq kg^{-1}	1.12	a, b, j
Grain	MJ kg^{-1}	15.2	c			
Diesel consumption	MJ L^{-1}	44.8	a, b, d, e	$kgCO_2$-eq. MJ^{-1}	0.08	a, b, j
Machine production	MJ L^{-1}	15.6	d, e, f			
Nitrogen (N)	MJ kg^{-1}	58.7	a, b, g	$kgCO_2$-eq kg^{-1}	5.68	a, b, j
P_2O_5	MJ kg^{-1}	17.1	a, b, g	$kgCO_2$-eq kg^{-1}	1.09	a, b, j
K_2O	MJ kg^{-1}	8.83	a, b, g	$kgCO_2$-eq kg^{-1}	0.52	a, b, j
Herbicide	MJ kg^{-1}	354	a, b, h, i	$kgCO_2$-eq kg^{-1}	23.3	a, b, j

a: Ecoinvent (2017), b: SIMAPRO (2017), c: Pimentel and Pimentel (2008), d: Bowers (1992), e: Richard (1992), f: Dalgaard et al. (2001), g: Kool et al. (2012), h: Mudahar and Hignett (1987), i: Grassini and Cassman (2011), j: IPCC (2013)

of rice straw off-field activities, as well as conversion factors, were adapted from the literature and databases from the SIMAPRO software.

The energy and GHGE conversion factors for agronomic inputs, processes, and products are presented in Table 10.2. The energy value and GHGE conversion factors of related materials were based on Ecoinvent database 3.0, which is one of the databases included in SIMAPRO software (Ecoinvent 2017), global warming potential over a period of 100 years (GWP-100a) of IPCC (2013) incorporated in SIMAPRO software (SIMAPRO 2017). The energy conversion for each manual labor agronomic activity conducted in rice production was adapted from Quilty et al. (2014). Global warming factors–100 years (GWP-100a) of CH_4 and N_2O were 30.5 and 265 kg CO_2-eq., respectively.

Production inventory data of energy and GHGE per unit of fertilizer chemicals refer to 1 kg N in urea ammonium nitrate with an N-content of 32%; 1 kg P_2O_5 in ammonium nitrate phosphate with a N-content of 8.4% and a P_2O_5-content of 52%; and 1 kg K_2O in potassium chloride with a K_2O-content of 60%. These data take into account production activities including transport of raw materials and intermediate products but do not account for waste treatment of catalysts, coating, and packaging. Similarly, energy and emission factors of herbicides are accounted for during their life cycle during production. Energy consumption and GHGE of machines were calculated based on 44.8 MJ L^{-1} of diesel (Ecoinvent 2017) accounted for production, transportation, and combustion in machinery. In addition to that, this value was added with 15 MJ L^{-1} for machine production (Bowers 1992; Dalgaard et al. 2001).

One of the findings from this study shows a comparison among the different rice-straw management scenarios (Fig. 10.7). Results show that incorporation of rice straw in the soil causes the highest GHGE whereas removal of rice straw reduces this impact significantly. Burning rice straw in the field causes not only high GHGE but also the highest human toxicology impact. Moreover, this burning scenario has the lowest net energy balance as it causes all the N contained in rice straw to be lost during burning. The study illustrates that rice straw removal from the field for purposes of mushroom or bioenergy production can effectively improve energy efficiency and reduce the environmental footprint of irrigated lowland rice production in Southeast Asia where straw burning is commonly practiced.

However, the presented data were obtained from a two-season experiment at a specific area in the Philippines, and thus might have limited scope for conclusions on national and global scales. Additional data from other regions or long-term

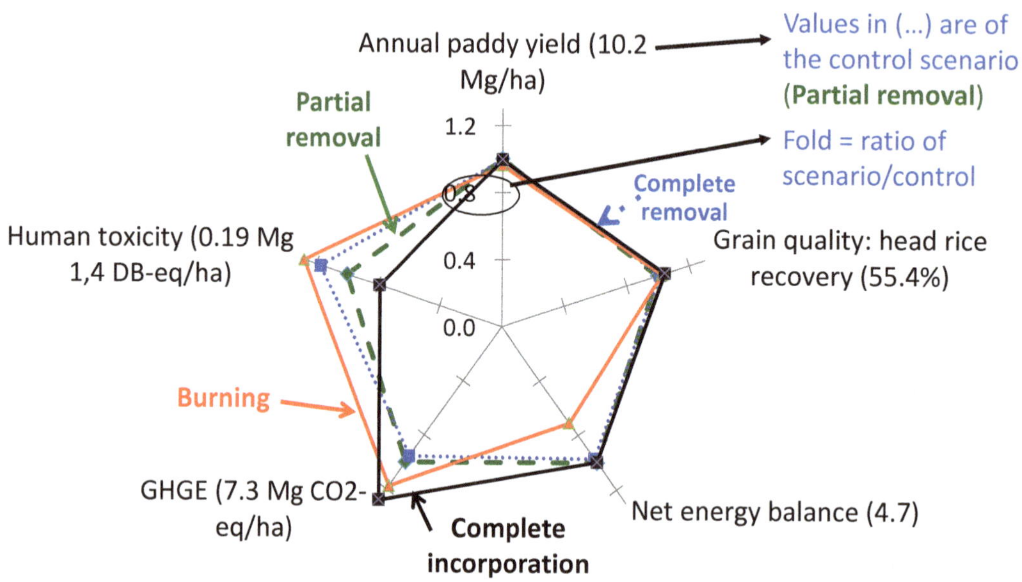

Fig. 10.7 Comparison of different rice-straw management scenarios. (Adapted from Nguyen et al. 2019)

experiments should be gathered and more utilization options of rice straw, such as for production of bio-char, compost, cattle fodder, bio-board, or bio-plastic, should be included in the LCA for a more comprehensive picture of the environmental footprint of different straw management alternatives.

10.5 Summary and Suggestions for Further Applications

LCA is used globally, in such programs as the Sustainability Consortium, ISO, UNEP, and others. In rice science, LCA should be used comprehensively to identify best practices of sustainable rice production, postharvest management, and rice-straw management. Energy balances, GHGE balances, and ecological and environmental impacts can be analyzed by using LCA and SIMAPRO. Internationally certified and reliable data for calculating energy and impacts are available in Agri-footprint, GHG protocol, Ecoinvent, etc., all incorporated in SIMAPRO.

References

Agri-footprint (2015) Agri-footprint 2.0 Part 1: Methodology and basic principles. Manual. Retrieved September 17, 2018 from www.agri-footprint.com

Bowers W (1992) Agricultural field equipment. In: Fluck RC (ed) Energy in farm production, vol 6. Energ World Agric, New York, pp 117–129

Brodt S, Kendall A, Mohammadi Y, Arslan A, Yuan J, Lee IS, Linquist B (2014) Life cycle greenhouse gas emissions in California rice production. Field Crops Res 169:89–98

Budavari Z, Szalay Z, Emi NB, Malmqvist T, Kth BP, Armines IZ, Circe GK, Sintef CW, Cai X, Calcon HS, Tritthart W (2011) Indicators and weighting systems, including normalisation of environmental profiles. Retrieved January 25, 2019 from https://www.sintef.no/globalassets/project/lore-lca/deliverables/lore-lca-wp5-d5.1-emi_final.pdf

Carrijo DR, Lundy ME, Linquist BA (2017) Rice yields and water use under alternate wetting and drying irrigation: a meta-analysis. Field Crops Res 203:173–180

Consultants Pr (2013) SimaPro LCA Software pre product ecology consultants. Amersfoot, Netherlands

Dalgaard T, Halberg N, Porter JR (2001) A model for fossil energy use in Danish agriculture used to compare organic and conventional farming. Agric Ecosyst Environ 87:51–65

De-Schryver AM, Brakkee KW, Goedkoop MJ, Huijbregts MAJ (2009) Characterization factors for global warming in life cycle assessment based on damages to humans and ecosystems. Environ Sci Technol 43(6):1689–1695

Ecoinvent (2017) Implementation of Ecoinvent 3. Retrieved January 25, 2019 from: http://www.ecoinvent.org/partners/resellers/implementation-of-ecoinvent-3/implementation-of-ecoinvent-3.html

European Commission Joint Research Centre (2009) International Reference Life Cycle Data System (ILCD) Handbook: specific guidance document for gener-ic or average Life Cycle Inventory (LCI) data sets. Draft for public consultation 01 June 2009. jrc.ec.europa.eu

Fusi A, Bacenetti J, González GS, Vercesi A, Bocchi S, Fiala M (2014) Environmental profile of paddy rice cultivation with different straw management. Sci Total Environ 494–495:119–128

Goedkoop M, Spriensma R (2001) The eco-indicator 99, a damage oriented method for life cycle impact assessment. Pre-Consultant. https://www.pre-sustainability.com/download/EI99_annexe_v3.pdf

Grassini P, Cassman KG (2011) High-yield maize with large net energy yield and small global warming intensity. Proc Natl Acad Sci U S A 109(4):1074–1079

Guinée JB (ed) (2004) Handbook on life cycle assessment operational guide to the ISO standards. Kluwer Academic Publishers, New York

Guinée JB, Gorrée M, Heijungs R, Huppes G, Kleijn R, Koning A, Oers, L, Sleeswijk AW, Suh S, Udo de Haes HA, Bruijn H, Duin R, Huijbregts MAJ (2001) LCA – An operational guide to the ISO standards, Final report

IPCC (2006) IPCC Guidelines for national greenhouse gas emission inventory,. Retrieved April 9, 2019 from: https://www.ipcc-nggip.iges.or.jp/support/Primer_2006GLs.pdf

IPCC (2013) Emissions Factor Database. Retrieved January 25, 2019 from: http://www.ghgprotocol.org/Third-Party-Databases/IPCC-Emissions-Factor-Database

Jeroen BG, Gjalt H, Reinout H (2001) Developing an LCA guide for decision support. Retrieved April 9, 2019 from: https://www.emeraldinsight.com/doi/abs/10.1108/09566160110392416

Kami DM, Barz M, Gheewala SH, Sajjakulnukit B (2012) Environmental and socioeconomic feasibility assessment of rice straw conversion to power and ethanol in Thailand. J Clean Prod 37:29–41

Kool A, Marinussen M, Blonk H (2012) GHG Emissions of N, P, and K fertilizer production, in LCI data for the calculation tool Feedprint for greenhouse gas emissions of feed production and utilization. Retrieved January 25, 2019 from: http://www.blonkconsultants.nl/wp-content/uploads/2016/06/fertilizer_production-D03.pdf

Linquist BA, Anders MM, Adviento-Borbe MA, Chaney RL, Nalley LL, da Rosa EF, van Kessel C (2015) Reducing greenhouse gas emissions, water use, and grain arsenic levels in rice systems. Glob Change Biol 21:407–417

Mudahar MS, Hignett TP (1987) Energy requirements, technology, and resources in the fertilizer sector. In: Helsel ZR (ed.), Energy in World Agriculture 2:26–61

Nguyen VH, Topno S, Balingbing C, Nguyen VCN, Roder M, Quilty J, Jamieson C, Thornley P, Gummert M (2016) Generating a positive energy balance from using rice straw for anaerobic digestion. Energy Rep 2016(2):117–122

Nguyen VH, Nguyen DC, Tran VT, Hau DH, Gummert M (2017) Energy efficiency, greenhouse gas emissions, and cost of rice straw collection in the Mekong River Delta of Vietnam. Field Crops Res 198:16–22

Nguyen VH, Sander BO, Quilty J, Balingbing C, Castalone AG, Romasanta R, Alberto MCR, Sandro J, Jamieson C, Gummert M (2019) An assessment of irrigated rice production energy efficiency and environmental footprint with in-field and off-field rice straw management practices. Energy Efficiency Journal, Springer (under review)

Pimentel D, Pimentel M (2008) Food, energy and society, 3rd edn. CRC Press, Boca Raton, FL

Quilty RJ, McKinley J, Pede VO, Buresh RJ, Correa JTQ, Sandro J (2014) Energy efficiency of rice production in farmers' fields and intensively cropped research fields in the Philippines. Field Crops Res. 168:8–18

Richard CF (1992) Chapter 5, energy analysis for agricultural systems. Energ World Agric 6:45–52

Shafie SM, Masjuki HH, Mahlia TMI (2014) Life cycle assessment of rice straw-based power generation in Malaysia. Energy 70:401–410

Shie JL, Chang CY, Chen CS, Shaw DG, Chen YH, Kuan WH, Ma HK (2011) Energy life cycle assessment of rice straw bio-energy derived from potential gasification technologies. Bioresour Technol 102:6735–6741

Silalertruksa T, Gheewala SH (2013) A comparative LCA of rice straw utilization for fuels and fertilizer in Thailand. Bioresour Technol 150:412–419

SIMAPRO (2017) SIMAPRO – LCA software. https://www.pre-sustainability.com/sustainability-consulting/sustainable-practices/custom-sustainability-software

Singh A, Pant D, Olsen SI (eds) (2013) Life cycle assessment of renewable energy sources. Green energy and technology. Springer, London. https://doi.org/10.1007/978-1-4471-5364-1_6. https://www.springer.com/gp/book/9781447153634

Suramaythangkoor T, Gheewala SH (2011) Implementability of rice straw utilization and greenhouse gas emission reductions for heat and power in Thailand. Waste Biomass Valor 2:133–147

US EPA (United States Environmental Protection Agency) (2006) Life cycle assessment: principles and practice. Retrieved January 20, 2019 from https://nepis.epa.gov/Exe/ZyPDF.cgi/P1000L86.PDF?Dockey=P1000L86.PDF

Chapter 11
Rice Straw Value Chains and Case Study on Straw Mushroom in Vietnam's Mekong River Delta

Matty Demont, Thi Thanh Truc Ngo, Nguyen Van Hung, Giang Phuong Duong, Toàn Minh Dương, Hinh The Nguyen, Ninh Thai Hoang, Marie Claire Custodio, Reianne Quilloy, and Martin Gummert

Abstract Rice straw is a tradable commodity in food and feed markets, particularly in rice-producing countries such as India, Vietnam, and Cambodia. Understanding the bottlenecks and linkages of different components and actors in rice and rice straw value chains is important to identify strategies to extract maximum value out of the straw and encourage diversion of straw utilization from unsustainable practices, such as burning, to more sustainable uses of rice straw. In Vietnam, for example, demand for mushroom and dairy products generate a market for high-quality straw, which can be an input for both industries. Mechanized straw collection is critical to supplying the byproduct as an input for these markets. Generally, the successful development of rice straw value chains will hinge on investments in intersectoral upgrading, triggered by the demand for food, feed, energy, and fiber from a growing urban population and the expanding food and nonfood industries. This chapter provides: (1) an overview of rice straw value chains, (2) a case study of rice straw mushroom value chains in Vietnam, and (3) suggestions for further developments.

M. Demont (✉) · N. V. Hung · M. C. Custodio · R. Quilloy · M. Gummert
International Rice Research Institute (IRRI), Los Baños, Laguna, Philippines
e-mail: m.demont@irri.org; hung.nguyen@irri.org; m.custodio@irri.org; r.quilloy@irri.org; m.gummert@irri.org

T. T. T. Ngo · T. M. Dương
Can Tho University, Can Tho, Vietnam
e-mail: ntttruc@ctu.edu.vn

G. P. Duong
Food and Agriculture Organization of the United Nations (FAO), Rome, Italy

H. T. Nguyen · N. T. Hoang
Ministry of Agriculture and Rural Development (MARD), Hanoi, Vietnam

M. Gummert et al. (eds.), *Sustainable Rice Straw Management*,
https://doi.org/10.1007/978-3-030-32373-8_11

Keywords Rice straw · Value chain · Intersectoral upgrading · Straw mushroom · Vietnam

11.1 Introduction

Value chains are by definition demand-driven (FAO 2014; Kaplinsky and Morris 2000). However, rice straw value chains—straw being a byproduct of rice value chains—are supply-driven rather than demand-driven. Rice straw is produced to satisfy the demand, not for rice straw but for rice, which drives rice value chain operations, from which rice straw is generated. This generation of the straw, in turn, triggers the need for proper use and management of the byproduct and, thus, the evolvement of straw supply chains. The demand for rice straw products in diverse markets will not necessarily trigger the production of the straw—as in a classic value chain—but will rather encourage the diversion of straw utilization from one activity (e.g., burning, incorporation) to another (e.g., baling, selling). The notion of an "end-market" is central to food value chain research (FAO 2014). In the case of rice straw value chains, the end-product is not food for human, but animal feed or an input into other food or nonfood value chains (e.g., dairy, meat, mushroom, energy, fiber, etc.). As a result, the "end-markets" for rice straw value chains are generally input markets for other food or nonfood value chains.

Value chain development typically follows an upgrading trajectory, which begins with process upgrading, moves on to product upgrading, and then on to functional, channel and intersectoral upgrading (Gereffi 1999; Kaplinsky and Morris 2000). Value chain upgrading—from process up to intersectoral upgrading—is triggered by demand factors (e.g., increasing demand for organic products) and then operationalized and reinforced by supply factors in response to the changes in the demand side (e.g., farmers' adoption of organic farming and traders' assurance of product quality and traceability throughout subsequent value chain stages). Urbanization, rising income levels, and diet change drive up the demand for high-value food products (Wang et al. 2014) and trigger a shift in food expenditure components, with a decreasing share of rice and rising shares of meat/fish, vegetables, and edible oils (Reardon 2015). As a result, actors on the supply side of rice value chains increasingly explore ways to offer customers more attractive rice products with superior quality attributes in terms of fragrance, purity, homogeneity, packaging, food safety, traceability, nutrition, health, and convenience. Rice businesses increasingly compete not just on price but on quality and product differentiation (Reardon et al. 2014). This triggers investment in process upgrading (e.g., adoption of postharvest and processing technology, good agricultural practices), product upgrading (e.g., improvement of intrinsic quality and extrinsic quality cues such as packaging, branding, and advertising), and functional upgrading (e.g., millers taking on the role of prefinancing quality inputs for farmers through contract farming to ensure reliable sourcing of quality paddy). Rising income levels also stimulate a shift towards consuming more processed products and prepared foods bought outside the home (Reardon et al. 2014; Reardon 2015). Together with the development of the industrial and manufacturing sectors and their concomitant demand for energy and fiber, this provides tremendous opportunities for intersectoral upgrading of rice value chains (Nguyen et al. 2017). Intersectoral upgrading of rice value chains may provide alternative income opportunities for value chain actors by diversification of income-generating activities and by adding value to existing rice products and/or byproducts (e.g., rice straw, rice husks, and rice bran). It also increases the sustainability of rice value chains by providing alternatives to existing unsustainable uses of byproducts, such as the burning of rice straw.

11.2 Mapping Rice Straw Value Chains in Vietnam, Cambodia, and the Philippines

Figure 11.1 shows a schematic diagram that maps rice straw value chains based on a multi-stakeholder workshop series conducted in Vietnam, Cambodia, and the Philippines under the BMZ-IRRI rice straw management project (IRRI 2019).

Rice straw value chains consist of three main functions: (1) harvesting, (2) collection and pretreatment, and (3) rice straw-based production and rice straw markets. It is worth noting again that, unlike food products whose end-markets are food consumers (via national and international food markets), rice straw generally ends up being used as an input for the production of other food and nonfood products. The end-markets of rice straw, as such, are often input markets. Currently, in the studied countries, rice straw is mainly used in the production of mushroom, ruminant feed, compost, mulching, and lining materials for the transportation of fragile fruits (e.g., watermelon) due to concrete, significant market demand for these straw-based products. Even though other new technologies and utilizations of rice straw (e.g., biochar, briquettes, bioenergy, biodegradable products, biofiber, etc.) are also already introduced, markets barely exist for these products (although they are emerging). Therefore, the markets for these relatively new straw-based products are categorized as "future markets".

The overall rice straw value chain process involves the collection of in-field rice straw and then transportation of the straw to end-users for their straw-based production activities. Specifically, rice straw that remains on the field after rice harvesting is collected, usually by contract service providers (who collect straw for a fee paid by rice farmers) or straw traders (who purchase in-field straw from farmers and then collect straw at their costs). Rice straw collection involves physical pretreatment of the straw, i.e., transforming straw that is spread around the field into various forms such as loose, baled, compacted, or chopped straw, depending on the specific subsequent uses.

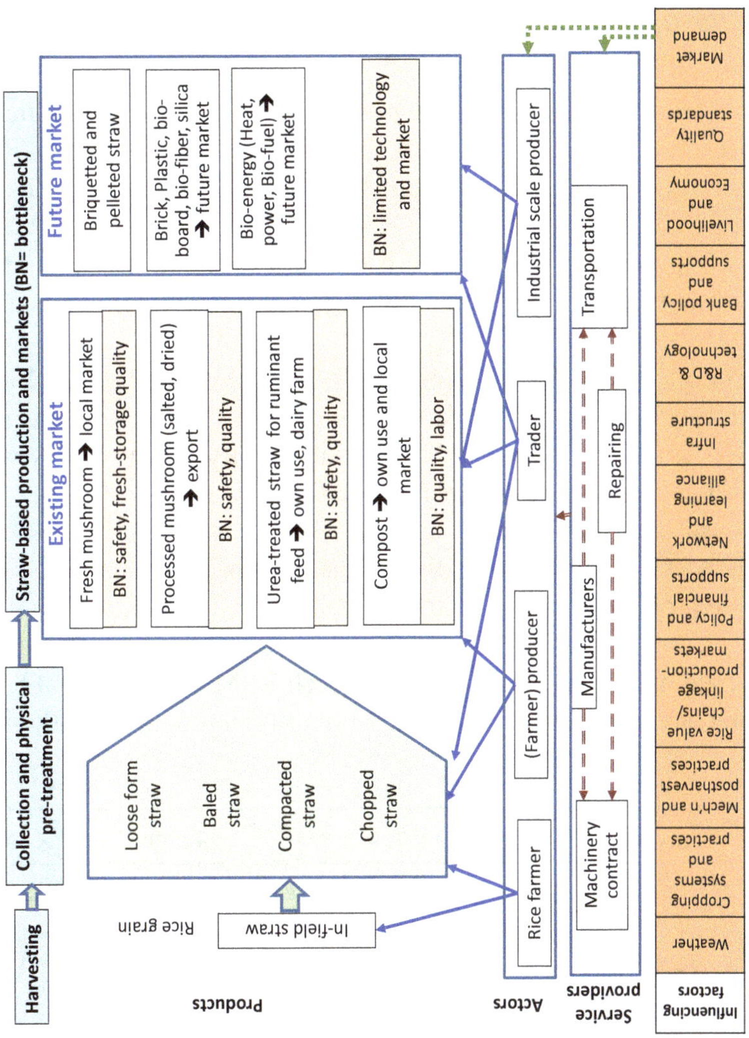

Fig. 11.1 Mapping of rice straw value chains

Rice straw value chains for the three countries are affected by many influencing factors, such as climate variability, rice variety, cropping systems employed, policies, technology availability, financial support systems, and extension systems.

11.3 Case Study of Rice Straw Mushroom Value Chains in Vietnam's Mekong River Delta (MRD)

11.3.1 Mapping Value Chains

Rice straw mushroom (*Volvariella volvacea*) (RSM) has been produced in the MRD for more than 20 years (Truc et al. 2013). However, the utilization of rice straw for mushroom production is still limited, using less than 5% of the total rice straw produced in the MRD. Can Tho City (CTC) and Dong Thap Province are most advanced in producing RSM. RSM is consumed both in fresh and processed forms, as illustrated through two channels targeting domestic markets and export markets (Fig. 11.2).

The domestic market of fresh mushroom mainly exists in the local areas, i.e., neighboring areas where mushroom farmers live. About 40–60% of this product is consumed in Ho Chi Minh City (HCMC), the largest city in the country with about 8.5 million residents, and nearby areas (100–300 km away).

As fresh mushroom can only be kept for 3 days (see Chap. 6), the amount (30–40% of the total RSM produced) that remains after being supplied to the fresh markets are then processed for export to China, Taiwan, Hong Kong, United States, and Europe.

In the domestic market, end-consumers of fresh RSM are individual households and institutional buyers. They buy fresh RSM mainly in the traditional or local wet

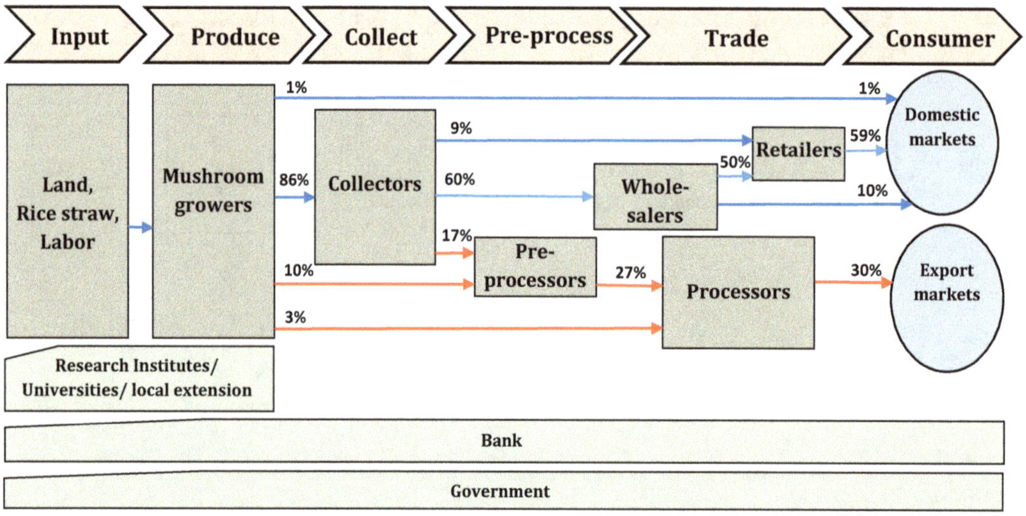

Fig. 11.2 RSM value chains in the MRD

markets. Less than 5% of the fresh product is consumed through supermarkets in Vietnam. About 80 to 90% of the fresh product is consumed by individual households who source mushrooms from local wet markets and supermarkets. The remaining 10% to 20% of RSM are consumed by institutional buyers (including staff and student canteens) and in restaurants. Institutional mushroom consumers buy the fresh product from both wholesalers and retailers.

11.3.1.1 Distributors

These actors include mushroom collectors and transporters, wholesalers, and retailers. RSM distributed to HCMC and some nearby markets are collected by collectors or transporters at farms of the growers or at the CTC assembly market.

11.3.1.2 Collectors and Transporters

These actors buy fresh mushroom from the production sites of mushroom growers or at the assembly market; they then deliver the product to preprocessors. The raw and fresh RSM includes cleaned and boiled products.

11.3.1.3 Pre-processors

The main roles of these actors are to classify the RSM (if they buy fresh from the growers) and then boil and salt or brine it. They deliver salted RSM at different sizes or grades, depending on the processors' requirements. There are reportedly six pre-processors in CTC and more than 10 in Dong Thap Province. Pre-processors invest in equipment and materials and buy the mushroom fresh, especially when the price is low.

11.3.1.4 Processors

There are two main processing companies in the MRD including Tu Thao Ltd., located in Soc Trang Province, and Quoc Thao Ltd. located in Vinh Long province. Their processed-RSM products are mainly exported to Asia, United States, and Europe. Products for exporting need to meet the required safety criteria set by the export markets.

11.3.1.5 Wholesalers

These actors are both primary wholesalers (big collectors near the production sites or assembly market) and secondary wholesalers in local markets in the MRD and wholesale markets in HCMC. Binh Dien wholesale market, one of the biggest wholesale markets in HCMC, consumes 50–60% of the total fresh RSM produced in the MRD. Thus, they have an important role in setting the market price and shaping demand for fresh RSM in the MRD. Lack of proper transportation and improper storage, poor transportation systems, and prolonged travel time usually deteriorate the quality of mushrooms arriving in wholesale markets and retailer markets in the MRD and HCMC. As a result, wholesalers usually sell all fresh RSM within the day.

11.3.1.6 Retailers

Retailers buy fresh RSM from primary or secondary wholesalers or collectors. The retailers operate in both wet markets and supermarkets but they handle a much larger amount of RSM in the former. Retailers also sell different types of vegetables including RSM at their wet market outlets. One risk in mushroom trading is that all fresh mushrooms must be sold within a day of reaching the market since quality will be lost given the poor preservation conditions.

11.3.1.7 Input Suppliers

These actors mainly include rice straw and spore suppliers. Labor and land are major inputs for RSM production. Labor includes both permanent and seasonal workers. The mushrooms are grown on land owned by growers or rented from other villagers such as rice land, free spaces in fruit gardens, or unused land in the community.

11.3.1.8 Rice Straw Collectors and Traders

Rice straw is collected either mechanically or manually. Mechanized collection is a good business model recently developed in the MRD (Nguyen et al. 2016). However, there is still a poor match between rice straw suppliers and mushroom growers in terms of specific mushroom quality characteristics, such as nutrient content (straw collected right after harvest) and minimal contamination as mentioned in Chap. 6.

11.3.1.9 Spore Suppliers

The spore business is generally a household enterprise. There are spore multipliers (suppliers) and spore agents. The agents buy spores from the suppliers who deliver them to the mushroom growers. There are perhaps as many as four spore suppliers in CTC and Dong Thap. They multiply spores (second or third generation) from pure spores. The challenges for suppliers are a general lack of (1) pure spore sources and (2) techniques to maintain spore quality, especially during the rainy season.

11.3.1.10 RSM Production

RSMs are produced by both rice farmers and mushroom growers. Some rice farmers make use of their own rice straw to produce the mushrooms. They often produce up to three mushroom cycles per year after harvesting rice. Mushroom growers are farmers who grow as many as six or seven RSM cycles annually. Outdoor mushroom growing (see Chap. 6) prevails in the area; little indoor growing is being done. Consumers prefer white-colored over dark-colored mushrooms. For producing white RSM, the growing beds require higher quantities of rice straw, which increases costs by 20–30%.

11.3.1.11 External Agents and Remaining Knowledge Gaps

These actors include local authorities, extension specialists with the Department of Agricultural and Rural Development and universities who provide services by introducing improved RSM production practices, and other institutes that conduct research. In addition, banks give loans to mushroom growers and other agents of the mushroom subsectors. Unfortunately, the contributions of these enablers to the development of RSM value chains in CTC and Dong Thap are insignificant. Even though the central government has recently issued regulations to support this subsector, insufficient research is being conducted on RSM value chains.

Thus, knowledge gaps remain in terms of improving RSM productivity, adding value, enhancing spore production and mushroom storage systems, and developing new RSM products. Technology transfer, mainly from the government, is very slow due to the lack of finance and the resistance to change among mushroom growers and other agents in the RSM subsector.

11.3.2 Economic Analysis

Figure 11.3 visualizes the annual farm-gate price trend of fresh RSM sold at harvest to fresh and processed-RSM markets in the MRD in 2015–2016 (Truc et al. 2017). Farm-gate prices are unstable due to annual fluctuations in demand generated by

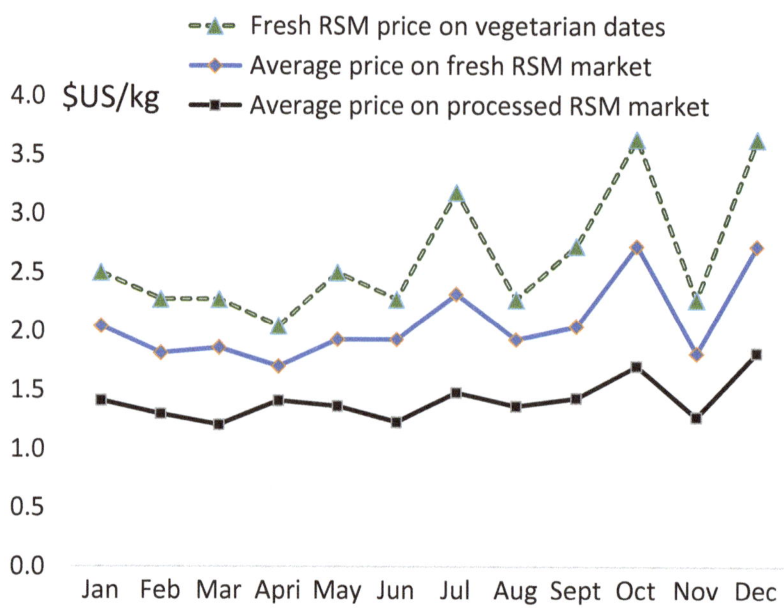

Fig. 11.3 RSM price at the farm-gate in the MRD in 2015–2016

vegetarian days and holidays, and because fresh mushrooms have to be sold within the day, as already explained.

The farm-gate price of fresh RSM on vegetarian days was about 2.0–3.6 $US kg^{-1} (40–120%) higher than on non-vegetarian days. Fresh RSM for the processed market usually has lower quality in terms of color, maturity, and size compared to the fresh-RSM market, resulting in price discounts. In addition, the processed-RSM market usually supplies higher quantities on non- vegetarian days when the demand for fresh RSM is lower than on vegetarian days. The farm-gate price for processed RSM is much lower, ranging between 1.2 and 1.8 $US kg^{-1} on average. The demand for vegetarian food consumption is higher in July, October, and December, driving up the price of fresh RSM during these times.

The significant gap (1.5–2.8 times) between the prices on fresh-RSM and processed-RSM markets provides mushroom growers with an incentive to schedule their RSM growing and harvesting activities according to the demand on the fresh-RSM market. However, as the quality of fresh RSM can degrade rapidly, it can only be sold in the domestic market within a day. The unsold fresh RSM (about 30% to 40% of total production) is then distributed to the companies that process and export the processed products.

Figures 11.4 and 11.5 visualize the value-added generated at different functions (stages) by the corresponding actors along the fresh and processed-RSM value chains and the selling prices of RSM at each stage. The value-added at a certain stage is the difference between the selling price (i.e., the price at which the product is sold to the actor of the next stage) and the intermediate costs (i.e., the cost of buying the product from the previous stage plus the cost of inputs incurred at this stage).

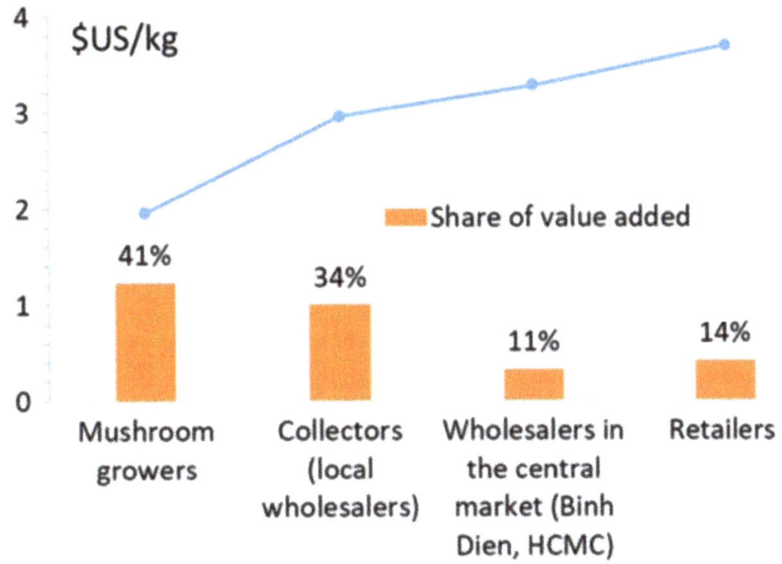

Fig. 11.4 Shares of value-added and selling prices by actors along fresh RSM value chains in MRD, Vietnam in 2015–2016. (Adapted from Toan (2018) and Truc et al. (2017))

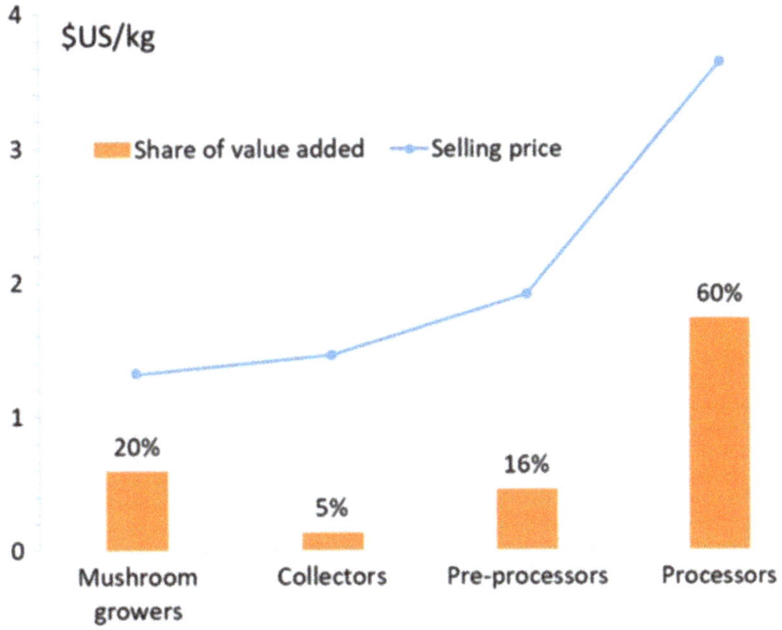

Fig. 11.5 Shares of value-added and selling prices by actors along processed-RSM value chains in MRD, Vietnam, in 2015–2016. (Adapted from Toan (2018) and Truc et al. (2017))

The contributions (shares) of actors along two RSM value chains to the total value-added are presented as percentages (orange bars in the two figures).

For fresh-RSM value chains, over one-third of the value-added is generated by the mushroom growers (1.20 US\$ kg^{-1}, representing 41% of the total share of value-added of fresh RSM). The remaining share of value-added mainly derives from transportation of RSM by the collectors and local wholesalers. However, fresh mushrooms are transported from production sites to the end users (located 100–300 km away) using simple preservation techniques, if any. As a result, fresh mushrooms degrade very fast (and should be consumed within a day). For processed RSM, processors are the actors who contribute and capture the largest share (75%) of the total value-added.

Although mushroom farmers generate high value-added in fresh-RSM value chains, they are often exposed to high levels of risk due to unstable yields and selling prices of fresh mushroom. Between 15% and 45% of RSM growers in Can Tho, Dong Thap and Hau Giang provinces in the MRD experience losses (Truc et al. 2017). The main risks in RSM production are unfavorable weather (for outdoor RSM growing), spawn contamination, and straw and water quality (see more details in Chap. 6).

The same case applies to processors, who contribute and capture the highest share of the value-added in processed-RSM value chains. They are also the only actors directly exposed to various risks associated with exporting RSM. The main challenge in their business is to ensure quality and traceability of processed RSM to comply with exporters' requirements. However, as processors procure fresh and preprocessed mushroom in open markets from collectors and preprocessors, they are often unable to trace inputs and mushroom sources, which are among the most important factors in quality control. In addition, they face changing consumers' preferences as well as changes in technical requirements of exporters.

Wholesalers generate the lowest share of value-added (11%) in the fresh-RSM channel. However, wholesalers earn daily profits of around US$ 1000–2000, while RSM growers make less than US$ 100 (which amounts to approximately US$ 6000 for each two-month cycle of mushroom growing); some even experience financial losses.

The significant profit gap between wholesalers and growers is caused by (1) the high perishability of fresh RSM (putting the growers under the pressure of having to sell the product as quickly as possible, even at lower prices); (2) the economies of scale for wholesalers who trade in bulk, as opposed to individual growers who sell mushrooms in small quantities; and (3) the larger markets that wholesalers can access (e.g., big cities), as opposed to farmers with limited market access.

11.3.3 Stakeholder Analysis

Stakeholder analysis reveals the linkages, relationships, roles, and influence of RSM stakeholders in the subsector. To conduct stakeholder analysis, key agents from RSM value chains were first identified (maximum ten agents). Second, they were ranked in terms of their importance and influence (1 = least importance or influence and 10 = highest importance or influence). "Importance" relates to the priorities given to satisfying needs and interests of each stakeholder. It is also their position or importance in relation to other actors/stakeholders based on their power,

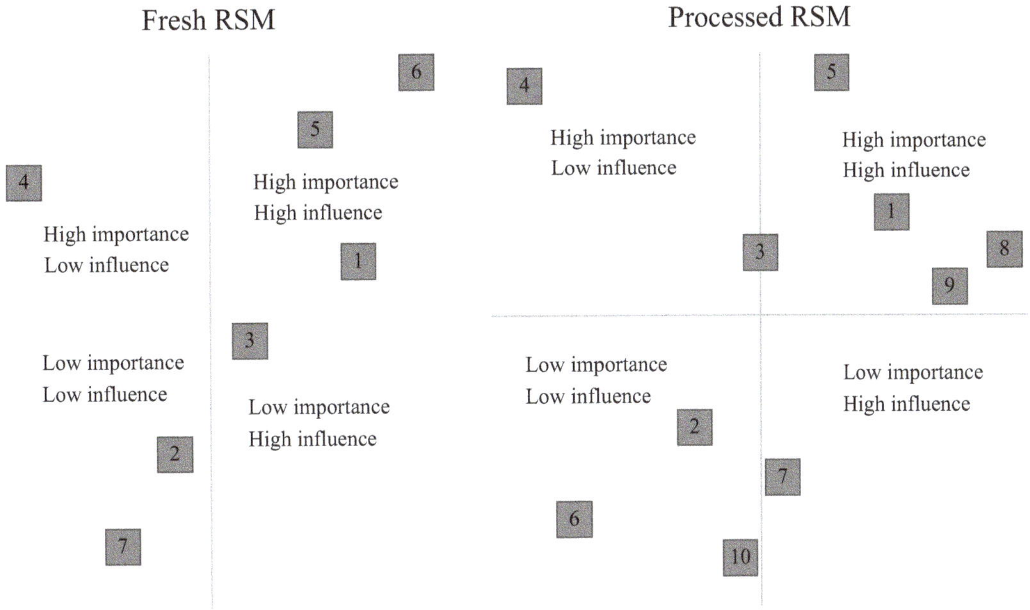

Fresh-RSM actors:
1. Mushroom growers
2. Spore suppliers
3. Rice straw suppliers
4. Collectors
5. Wholesalers
6. Transporters
7. Retailers
8. Individual consumers
9. Institutional consumers
10. Extension workers

Processed-RSM actors:
1. Mushroom growers
2. Spore suppliers
3. Rice straw suppliers
4. Collectors
5. Preprocessors
6. Processors
7. Extension workers

Fig. 11.6 Importance-influence matrix of RSM in the MRD

which is stipulated by trading capacity, quantity, quality, and price management in the current market. "Influence" relates to the power of a stakeholder to facilitate or impede the achievement of an activity's objective. In this analysis, the criterion used to evaluate the influence of the stakeholders was their contribution to the success of expanding the RSM subsectors (including increasing both volume and sales). Then, the ranked agents were mapped in the importance-influence matrix (Fig. 11.6).

The stakeholder analysis was first conducted by the research team using data from interviews of individual actors. Then, it was verified in a multi-stakeholder workshop in which importance and influence were assessed.

In the fresh-RSM market, wholesalers in Can Tho, Dong Thap, and Binh Dien are the most important players. They set or influence the price of mushroom on the market on a daily basis because they know who will consume the fresh product. Extension workers and transporters have the least important roles in both the fresh and processed-RSM value chains.

In order to expand RSM production for the fresh market, extension workers should engage more with other agents to improve productivity and value-added by speeding up the transfer of production, trading, and processing technologies. Transporters and collectors could increase their role and strengthen their links with mushroom growers and local wholesalers. Otherwise, transport should be integrated into wholesalers' business models leading to fewer intermediaries in the value chain with stronger linkages ("dis-intermediation"; see Reardon et al. 2014; Reardon 2015).

To expand the processed-RSM market, processors need to further explore the demand for processed products in the domestic and overseas markets. Moreover, the expansion of both the fresh and processed-RSM markets requires collaboration and linkages between agents to govern the traceability and quality of inputs (straw and spore), straw mushroom-growing practices (increase productivity and safety) and preprocessing and processing procedures (safe and good practices) to meet increasingly strict market requirements. More research on the demand of new products and consumer behavior in export markets will help expanding those markets.

11.3.4 Constraints of RSM Value Chains

11.3.4.1 Low Economic Returns and High Risks

The first constraints in RSM value chains are the low economic returns and high risks involved in the business. In the domestic fresh-RSM market, growers generate the highest value-added along the fresh-RSM value chain; however, they receive lower daily profits compared to other actors. Growers are applying traditional methods to produce RSM (open-field practices). This leads to low yields, high investment costs, and environmental risks. Fresh mushroom is not preserved properly during transportation from production sites to end-markets, leading to high damage and shorter shelf-life of products.

11.3.4.2 Lack of Linkages among Actors

This is one of the main challenges in most agricultural value chains in developing countries (Reardon et al. 2014; Reardon 2015). Processors are facing increasingly strict export requirements; but as they do not have reliable contract-based links with preprocessors, mushroom growers and input supplies, it is difficult for them to govern the traceability and quality of RSM production processes. From the perspective of the growers, since they are not strongly linked to input suppliers and traders, they are not well-informed about the characteristics and quality of the rice straw and the spawn that they use. Thus, it is difficult for growers to ensure quality inputs and, hence, govern product quality.

11.3.4.3 Little Support for Strengthening Capacity

There is little support from external actors to strengthen the capacity of internal actors in RSM value chains. The main external actors, including local extension officers, researchers and financial providers, are not active in conducting and accelerating RSM-related research, technology transfer, and financial support to the value chain actors.

11.3.5 SWOT Analysis

We conducted a SWOT analysis to identify upgrading strategies for RSM value chains and explore opportunities for expanding the value chain subsector to divert the use of available rice straw in the MRD from burning to more value-added income opportunities. Table 11.1 provides a summary of the analysis.

Table 11.1 SWOT analysis of RSM value chain in MRD

SWOT analysis	**Strengths**	**Weaknesses**
	1. Available straw in the MRD is abundant	1. Small-scale mushroom production and trading
	2. Available mushroom growers, input suppliers and buyers with experience	2. Low economic return and risky business
		3. Few linkages among RSM actors
		4. Low quality of input supplies
Opportunities	**S-O strategies**	**O-W strategies**
1. High demand for both fresh and processed mushroom from domestic and export markets	Expand the mushroom subsector in both size and quality	1. Improve the efficiency of the RSM subsector.
2. Support from the Vietnamese government to expand the RSM subsector		2. Improve linkages among mushroom actors to improve efficiency, quality, and scale
3. New technologies to improve the performance of RSM value chains		
Threats	**S-T strategies**	**W-T strategies**
1. Increasing technical restrictions from export markets	Improve the quality governance and efficiency of RSM value chains	Control and improve the quality of input supplies as well as mushroom products and value chain processes.
2. Higher cost of straw, labor and land		

11.3.5.1 Strengths

- Rice straw in the MRD is abundant without any specific utilization. About 40–80% of the rice straw in the MRD is still burned, especially in three rice cropping systems.
- Available mushroom growers, input suppliers and RSM buyers (straw and spore suppliers, transporters, wholesalers, retailers) with long experience working in the RSM subsector for 10–30 years. They will be valuable resources in supporting the expansion of the value chains.

11.3.5.2 Weaknesses

- Small-scale and individual mushroom production and trade that hinder mushroom growers, input suppliers and traders from engaging in large orders for mushroom products with specific and strict quality requirements.
- Low economic return and risky business due to low adoption of advanced techniques and technologies from production to transportation of fresh mushroom.
- Few linkages among RSM actors.
- Low quality of input supplies (straw and spores).

11.3.5.3 Opportunities

- High demand for RSM in both domestic and processed markets, especially the demand for safe, healthy and convenient products.
- The Vietnamese government supports the expansion of the RSM subsector and has issued policies and directions for its development.
- The availability of new technologies supporting the RSM subsector (production, transportation, and preservation).

11.3.5.4 Threats

- Increasingly strict technical requirements and standards in export markets as the latter have shifted from low-income (China, Taiwan, Hong Kong) to high-income markets (United States, Europe, Japan).
- Rising costs of straw, labor, and land.

11.3.5.5 Recommended Strategies to Upgrade RSM Value Chains in the MRD

From the SWOT analysis (Table 11.1), two recommended strategies to upgrade RSM value chains in the MRD are summarized below.

- Improve the efficiency of the mushroom subsector. Explore and adopt new technologies that can reduce costs of straw collection and spore and RSM production as well as product preservation and processing.
- Improve the linkages among RSM value chain stakeholders to upgrade and expand the value chains. The purpose of linkages is to improve the efficiency and ensure the quality of mushroom products. Both RSM quality and quantity should be governed. Wholesalers and exporters should take the leading roles in linking the growers and other input suppliers.

11.3.5.6 Recommended Prioritized Activities to Support Fresh and Processed Mushroom Value Chains

- Enhance value chain linkages between rice farmers (who supply straw), mushroom growers, supporting actors (input suppliers, labor), and external agents.
- Transfer indoor mushroom growing techniques and other improved techniques to growers to improve productivity and profit from RSM production.
- Conduct market research to explore new market opportunities for fresh and processed mushroom products in the domestic and export markets.

11.4 Summary, Further Research, and Developments

RSM value chains have been adapted and improved over the last 40 years. However, the profit margin for actors is still low and participation in RSM value chains involves a high level of risks. Mushroom growers are facing many difficulties, such as the lack of a reliable supply of good quality of straw and spawn. Mushroom yield is low and fluctuates (0–4% of rice straw use) due to the fact that it is mostly produced outdoors. Mushroom growers are still reluctant to shift and invest in improved mushroom production techniques and indoor models.

Speeding up technology expansion and controlling straw and spawn quality will significantly improve profits for mushroom growers. Enhancing the linkages and coordination along the value chains will improve their overall performance and foster upgrading. Improving fresh-mushroom transportation systems will enhance mushroom quality. Finally, sound market research to explore new opportunities for fresh and processed-mushroom products is needed to inform actors and crowd-in investment in value chain upgrading.

Value chains and influencing factors of rice straw in Vietnam, Cambodia, and the Philippines were mapped through multi-stakeholder workshops with experts from agriculture and the food and energy sectors. In terms of future markets for agricultural uses, Vietnamese and Cambodian stakeholders proposed organic fertilizer, crop mulching, and biochar, while Philippine stakeholders suggested rice straw nets/mats, rice straw gardens, and seedling pots and trays for mechanical rice transplanters. For future food and feed markets, stakeholders from all three countries agreed that animal feed is the most promising product. Cambodian stakeholders also saw potential in mushrooms as a future opportunity that needs to be developed. When looking at future markets for energy and industry, Vietnamese and Cambodian stakeholders identified bioplastics and biofuels as the most promising, whereas Philippine experts suggested biofilters/desiccants, nanomaterials, and textiles.

Investment in product and technology development and quality upgrading were identified to be the main upgrading strategies required to develop rice straw value chains in the three countries. Investment in rice straw value chain upgrading should be preceded by proper end-market analysis to assess consumer acceptance of the new products that are derived from rice straw.

The current rice straw "supply" chains observed in the three countries are short and supply-driven. All three are on a similar trajectory of intersectoral upgrading with Vietnam leading, closely followed by Cambodia and the Philippines lagging behind. However, Philippine stakeholders identified a great diversity of future end-markets and products. The rich data obtained through the stakeholder workshops in the three countries indicate that many market opportunities for rice straw are still untapped and more R&D will be needed to develop these opportunities. The more these markets are developed, the more rice straw will evolve into a commodity with increasing market value, and the more farmers will receive monetary (and nonmonetary) incentives to move away from straw burning towards more sustainable uses of rice straw.

References

FAO (2014) Developing sustainable food value chains: guiding principles. Food and Agriculture Organization of the United Nations (FAO), Rome. http://www.fao.org/3/a-i3953e.pdf

Gereffi G (1999) International trade and industrial upgrading in the apparel commodity chain. J Int Econ 48(1):37–70

IRRI (International Rice Research Institute) (2019) Final report of the BMZ-funded IRRI sustainable rice straw management project (unpubl.)

Kaplinsky R, Morris, M (2000) A handbook for value chain research (Vol. 113). University of Sussex, Institute of Development Studies

Nguyen VH, Nguyen CD, Van Tran, T, Hau HD, Nguyen NT, Gummert M (2016) Energy efficiency, greenhouse gas emissions, and cost of rice straw collection in the Mekong River Delta of Vietnam. Field Crop Res 198:16–22

Nguyen VH, Balingbing C, Quilty J, Sander BO, Demont M, Gummert M (2017) Processing rice straw and husks as coproducts. In: Sasaki T (ed) Achieving sustainable rice cultivation. Burleigh Dodds Science Publishing Limited, Cambridge, pp 1–33

Reardon T (2015) The hidden middle: the quiet revolution in the midstream of agrifood value chains in developing countries. Oxf Rev Econ Policy 31(1):45–63

Reardon T, Chen KZ, Minten B, Adriano L, Dao TA, Wang J, Das Gupta S (2014) The quiet revolution in Asia's rice value chains. Ann N Y Acad Sci 1331:106–118

Toan DM (2018) Mushroom value chain in Dong Thap Province, Vietnam. Masters thesis. Can Tho University. In: Vietnam

Truc NTT, Sumalde ZM, Palis FG, Wassmann REINER (2013) Farmers' awareness and factors affecting farmers' acceptance to grow straw mushroom in Mekong River Delta, Vietnam and Central Luzon, Philippines. Intl J Environ Rural Dev 4(2):179–184

Truc NTT, Toan DM, Hung PQ (2017) Including gender issues and value chain analysis to develop straw mushroom production in the Mekong River Delta: Case study Vinh Loi District, Bac Lieu Province. Research Report. CCAFS-IRRI

Wang H, Moustier P, Nguyen TTL (2014) Economic impact of direct marketing and contracts: the case of safe vegetable chains in northern Vietnam. Food Policy 47:13–23

www.ingramcontent.com/pod-product-compliance
Lightning Source LLC
LaVergne TN
LVHW082232200926
848484LV00008B/2346
9781013274619